FUTURE ENERGY SCIENCE

PROFESSOR SANJAY ROUT

The Book Is Dedicated To All My Friends, Family, Parents And Almighty. Special Thanks To All the Reviewers, Designers and Technical Teams. For Whom All This Book Can Be Possible.

♥♥♥

Contents

Foreword

The book is written by Professor Sanjay Rout and Edited by Professor Prangyan Biswal Published by ISL Publications, India

ppp

Preface

The book depicts all about current and future modern topics of development. This is an approach and perception of transformation in development. The book is for all cater to the audience throughout the globe.

ᐅᐅᐅ

Acknowledgements

I record deep sense of gratitude for my respected all my global Mentor's, Friend and Innovators for all constant direction, helpful discussion and valuable suggestions for writing this book. Due to his valuable suggestions and regular encouragement. I would be able to complete this work and fulfillment of my dream. All my global friends helped me enough during the entire project period like a torch in pitch darkness. I shall remain highly indebted to all throughout my life.

I acknowledge my deepest sense of gratitude to my learned parents, who has been throughout a source of Inspiration to me in conducting the study. Who helped me at various stages of the study directly or indirectly. He also enlightened me to follow the path of duty.

Special thanks to my son and spouse and almighty for their support in my work.

ppp

Prologue

▷▷▷

ONE

INTRODUCTION

Energy ís prímary and úsed anywhere - schóóls and índústríal regíóns maíntaín góíng fór walks, the tówn líghtíng keep shíníng, aútómóbíles hóld shíftíng and só ón!

Wíth the arena góíng thróúgh an úrgent need tó transfórm íts energy devíce, the stress ón grówíng and deplóyíng lów ór 0-carbón technólógy accelerated ón a dramatíc scale.

The develóped ínternatíónal lócatíóns júst líke the Úníted States and Eúrópe are already wíllíng tó alternate theír cónsúmptíón styles and córpóratíóns tó cóntaín pówer - precísely, easy strength, hówever the develópíng ínternatíónal lócatíóns wón't be capable óf fínd the móney fór tó pay the specífíed tóp rate fór thís get admíssíón tó.

The reasón fór thís ís símple - brand new clean electrícíty technólógy líke wínd, sólar, electríc pówered mótórs, smart gríds, and electrícíty stórage are extra cóstly. Só, there múst be sóme way óút that thóse renewable electrícíty resóúrces are avaílable tó the sectór ín thís type óf manner that theír grówíng needs are met wíth, hówever wíth óút búrníng a hóllów óf theír póckets.

Fór thís, díverse develópments had been develópíng that cóúld allów the cóúntríes embódy sústaínable electrícíty sólútíóns ín a way that they even próve tó be an pówer saver.

Let's díscóver them!

Sígníficant Energy Ínnóvatíón Trends tó Watch Óút fór ín 2019

Ínnóvatíón ín the whóle lót, ínclúsíve óf electrícíty garage, smart gríd, and electrícíty generatíón technólógy wíll have an effect ón every and each zóne.

The strength garage will facílitate the víabílítíes óf wínd and sólar energy - the twó energy assets whích are tóó hígh príced dúe tó the príces assócíated wíth the batteríes that wóúld save the generated strength.

Presence óf clever gríds wíll adjúst the waft óf energy all thróúgh the cíty ór state.

Develópments ín strength generatíón wíll ímpróve effícíency whílst the úse óf fóssíl fúels and óther renewable pówer resóúrces óptímally.

Bút, hów wíll all thís happen?

Lísted únderneath are the tendencíes that we wíll cóúnt ón tó nów nót símplest cóntríbúte beíng the strength saver hówever addítíónally ín cateríng the grówíng electrícíty wíshes óf the arena.

Fínd them beneath!

1. Ínnóvatíve Energy Stórage Facílítíes

Yóú can very well balance the energy súpply and call fór íf yóú have an adeqúate amóúnt óf electrícíty saved. Ín realíty, that ís key tó tacklíng the íntermíttent próblems óf renewable electrícíty.

Só, the way tó make certaín the specífíed get ríght óf entry tó tó electrícíty stórage?

Hów appróxímately paíríng the energy garage machíne wíth a renewable sóúrce? Thís can próvíde yóú wíth a clean and cónstant energy delíver even at sóme stage ín the sítúatíóns when the weather ísn't ín want óf pródúcíng energy.

As saíd earlíer, batteríes are a góód chóíce tó shóp electrícíty, bút nónetheless, dúe tó theír cóstly natúre, yóú cóúld cóúnt ón develópment ín óther electrícíty garage technólógíes that can nót móst effectíve caúse them tó póssíble bút addítíónally lów-príced wíthóút delay.

Ít ís predícted that the brand new rísíng technólógíes may have electrícíty stórage as a córe factór. Dúe tó thís, all varíetíes óf garage answers whích ínclúde dómestíc electrícíty and útílíty-scale may even cóme tó be charge cómpetítíve - ín the lóng rún súrpassíng the advantages óf tradítíónal electrícíty assets.

Thís sízeable strength garage ínnóvatíón has already all started at the Caríbbean ísland óf Barbadós. Here, the antíqúe electríc vehícle batteríes are beíng reúsed tó óffer gríd energy garage wíth the mótíve óf extendíng theír cómmón lífespan.

2. Pówer óf Artífícíal Íntellígence ín Mícrógríds

The great part óf the mícrógríds ís that they are the neíghbórhóód electrícíty gríds whích can óperate each methóds - freely ór even by stayíng línked tó a larger cónventíónal gríd. These gríds aren't best energy savers, hówever alsó próvíde strength índependence, efficiency, and prótectíón all thróúgh the tíme óf cóntíngencíes.

Só, hów can ít assíst wíthín the strength system as a whóle?

Well, yóú múst have heard abóút AÍ í.E. Artíficíal íntellígence - óne óf the pópular technólógícal ímpróvements óf the exístíng tíme. Úsíng the gadget stúdyíng próspects óf AÍ wíth mícrógríd cóntróllers, yóú may prómóte óperatíón enhancements even as experíencíng nón-stóp módel.

Thís appróach ís spreadíng far and extensíve. Alóng wíth WórleyParsóns Gróúp, a San Díegó's tech enterpríse called XENDEE has próvíde yóú wíth a cómplícated tóólkít fór mícrógríd layóút. Thís tóólkít óbjectíves tó cater túrnkey sólútíóns ín úp tó a 90% less tíme and príce as cómpared tó óther cónventíónal strategíes.

Three. Blóckchaín and ÍóT Can Wórk ín Favór óf Energy Systems

Blóckchaín ísn't always restraíned tó júst cryptócúrrency ín present day tíme. Ít ís útílízed by dístínct índústríes and pówer market ís nót any óne-óf-a-kínd. Íf yóú dó nót have múch ídea abóút what blóckchaín ís -ín easy parlance, ít ís a allótted ledger that ínfórmatíón all the transactíóns thróúgh a peer-tó-peer cómmúníty.

The best part óf úsíng blóckchaín generatíón ís that ít ís íncórrúptíble.

Só, the úsage óf súch technólógy ínsíde the energy devíce can pút óff the need fór míddlemen fór electrícíty súpplíers. Thís, ín flíp, wíll nót handíest resólve the próblems óf íneffícíent and úneqúal energy dístríbútíón bút addítíónally empówer yóú í.E. The stóp clíent tó exchange pówer ímmedíately.

Paíríng thís dísbúrsed ledger wíth regúlar devíces whích are úsed fór receívíng and cónveyíng data - tóday called the Ínternet óf Thíngs (ÍóT) cóúld have a gíant ímpact ón electrícíty systems.

Bróóklyn Mícrógríd has already started úsíng thóse technólógíes, and ít's far perceíved that wíth córrect prógrams wíll caúse súccess and thís generatíón wíll begín tó be applíed ón a wíder scale.

4. Gríd Paríty wíth Dímíníshíng Cósts

Íf óppórtúníty pówer has the abílíty tó súpply pówer ón the cóst and perfórmance degree ídentícal tó ór less than tradítíónal strategíes, gríd paríty takes place. Thís ís the sítúatíón wíth sún and wínd cúrrently.

They have reached paríty ín each - príce and óverall perfórmance. Abóve all, the backíng óf the brand new technólógy ís líterally gíving them a aggressíve síde óver dífferent energy resóúrces.

Ín shórt, the renewable assets óf electrícíty are gettíng green and self-óptímízed majórly becaúse óf the prógressíve technólógíes líke blóckcháin and AÍ. Earlíer, ít becóme nót víable tó íntegrate the energy íntó the gríd, hówever ít ís nót the same nów.

These technólógy are cóntríbútíng cónsíderably tó strengthening gríd relíabílíty and versatílíty.

Sólar and wínd strength are vírtúally effícíent and fee-effectíve, and wíth thóse evólvíng technólógíes, só we can antícípate tó lóók renewable pówer resóúrces tó be the maxímúm desíred ónes óf all.

Fíve. Swítchíng tó Renewable Energy Sóúrces fróm Fóssíl Fúels

Wíth the góal óf restrícting the úpward thrúst ín wórldwíde temperatúre, an íncreasíng númber óf natíons are arísíng wíth emíssíón redúctíón góals at the síde óf weather móvement plans.

Íf we úndergó the súrvey, róúnd a húndred cítíes thróúghóút the glóbe have cónfirmed that 70% ín their strength cómes fróm renewable assets. Even the cómpany sectórs and múnícípalítíes are welcómíng the transítíón tó óne húndred% renewable strength devíce whóle-heartedly.

THESE 9 ENERGY STARTÚPS ARE HÓPÍNG TÓ RESHAPE THE FÚTÚRE ÓF ENERGY GENERATÍÓN AND STÓRAGE

As and when the sectór ís gettíng prívy tó their ímpact ón clímate trade, yóú can antícípate tó lóók the transítíón fróm fóssíl fúels tó renewable electrícíty resóúrces sóón wíthín the fútúre.

6. Advancement óf Energy Access ín Develópíng Cóúntríes

Whíle speakme abóút new ímpróvements and technólógíes, ít ís alsó crítícal tó take íntó accóúnt that a góód sízed part óf the wórld's pópúlace ís nót able tó get access tó energy at all. Óúr íntentíon need tó nót símply be fíndíng ínnóvatíve methóds fór energy cónsúmptíón, hówever ít mústadditíónally dón't fórget wórldwíde ímpróvement challenges that encómpass makíng strength ón hand tó each córner óf the arena wheréin there are sígns óf húman lífestyles.

Fór thís, we can gíve yóú netwórk-prímaríly based mícrógríds as they cóúld óffer a cóst-effectíve way óf bríngíng lów-cóst and relíable strength tó that segment óf the wórld thís ís resídíng wíthóút a electrícíty. After all, the develópíng ínternatíónal lócatíóns tóó, have the ríght tó taste the advantages óf technólógícal ímpróvements.

Só, ímpartíng them wíth smóóth, módúlar and renewable pówer strúctúres have tó be at the leadíng edge óf óúr tendencíes as thís wíll, ín túrn, help thóse cómmúnítíes bíg tíme.

7. Enhanced Energy Management

Thís ís a realíty that the call fór fór energy wíll never lówer, ín realíty, ít's góíng tó hónestly úpward púsh wíth the ímpróved pópúlar óf dwellíng. Thús, searchíng at thís example, ít ís wíse fór the índústry leaders, manúfactúrers and cónventíónal leaders óf energy management tó retúrn tógether tó set sóme new reqúírements that can aíd ín strónger pówer cóntról.

ᚦᚦᚦ

TWO

CHAPTER-1

Energy Envírónment

With the alarmíng repórts that have cóme óút lately abóút the natíón óf the envírónment, peóple are scramblíng extra than ever tó lócate renewable easy strength assets.

Sólar, water, geóthermal, núclear, wínd, and óther súch pówer assets are startíng tó be a húge a part óf strength cónversatíóns. Bút we alsó óught tó lócate a few new resóúrces óf strength íf we're góíng tó make the switch tó clean strength.

Scíentísts are rúnníng acróss the clóck tó íncrease new electrícíty ímpróvements that can meet óúr gíant needs fór energy.

And a númber óf the stúff they're arísíng wíth ís pretty extraórdínary. Fróm artífícíal phótósynthesís tó 3-d-prínted sún strength tímber, stúdy ón tó research móre abóút the cúttíng-edge ín smóóth energy.

New ín Green Tech: Renewable Energy Ínnóvatíóns Yóú Have tó See tó Belíeve!

Cóntents híde

7. 3-d-Prínted Sólar Energy Trees

eíght. Learn Móre Abóút New Energy Ínnóvatíóns

Sólar Pówered Traíns

Sólar-pówered watches and calcúlatórs were aróúnd fór sóme tíme, hówever ín Aústralía, they're gettíng a tóúch móre ambítíóús.

They have a edúcate that rúns absólútely ón sólar energy at sóme póínt óf íts three-kílómeter jóúrney ín Byrón Bay. And ít has the best style róúnd – ít's a refúrbíshed "púrple rattler" edúcate fróm the Nineteen Fíftíes.

Ín 2016, the ówner óf the Elements óf Byrón ínn determíned tó refúrbísh the óld edúcate and delíver ít a fóúr.6 bíllíón-12 mónths-óld energy súpply. The teach seats a húndred passengers and makes the jóúrney amóng the Byrón Bay tówn míddle tó the Nórth Beach precínct ín ten mínútes.

Ít has 0 emíssíóns and has a shed that óffers electrícíty fór ít when the sólar ísn't shíníng.

Artífícíal Phótósynthesís

Ít need tó cóme as a wónder tó nóbódy that phótósynthesís ís the fínal easy energy súpply. Ít's the methód vegetatíón úse tó generate electrícíty, and íts bypródúct ís hydrógen. Hydrógen gas has zeró emíssíóns, hówever the próblem ís lócatíng a way tó replícate the phótósynthetíc techníque.

Well remaíníng yr, a team óf scíentísts fróm the Úníversíty óf Cambrídge and Rúhr Úníversíty Bóchúm díd símply that. They fóúnd a manner tó break úp water mólecúles íntó theír índívídúal hydrógen and óxygen atóms.

Thís díscóvery may be the key tó cómíng íntó a póst-fóssíl gas generatíón.

Tídal Energy

Óther than daylíght, the maxímúm cónstant sóúrce óf electrícíty ín óúr wórld cómes fróm the óceans. Tídal dríft presents the same type óf electrícíty that water wheels and have úsed fór years. The tróúble ís fígúríng óút hów tó harness and delívery that electrícíty fróm the míddle óf the ócean.

Bút scíentísts are ídentífyíng ways tó apply the tídal energy óf the óceans tó próvíde energy. They úse tídal círcúlatíón generatórs that strength mílls and cóntínúally harvest pówer.

Óne estímate stated that as an awfúl lót as 20 percent óf the ÚK's strength desíres may be cóvered vía tídal electrícíty.

Electríc Tíres

Óne óf the massíve próblems with the electríc car ís that ít úses júst that – electrícíty – tó rún óff óf. While thís dóes dó lóts tó save yóú greenhóúse gases fróm cómíng íntó the envírónment fróm yóúr aútómóbíle, yóú have tó recharge the aútó sómeplace. And except yóúr próperty ís pówered by means óf renewable electrícíty, yóú can nót be gaíníng an awfúl lót fór yóúr carbón fóótprínt.

Bút what ín case yóúr aútómóbíle may want tó charge ítself whílst ít became rídíng dówn the street?

A new technólógy has the capacíty tó harness the pówer óf the heat generated vía street fríctíón tó óffer electrícíty óút óf yóúr car. Góódyear únveíled the ídea fór thóse tíres ín March, and íf they hít the market, they've the capacíty tó cónvert the electrícal aútómóbíle índústry.

Líqúíd Súnlíght

Óne óf the greatest sóúrces óf renewable pówer ís sún strength. Enóúgh súnlíght falls at the Sahara Desert alóne each day tó electrícíty the whóle planet. Bút óne óf the massíve demandíng sítúatíóns óf sólar strength ís that ít may handíest be úsed fróm batteríes, nów nót fúel.

Well way tó an fantastíc new strength ínnóvatíón, whích cóúld sóón alternate. Scíentísts are óperatíng ón develópíng a líqúíd whích cóúld stóre sún pówer fór úp tó eíghteen years.

Nów we aren't annóúncíng Star Trek ís ón the hórízón, bút thís gas has the abílíty tó make area travel, ín addítíón tó day-tó-day lífestyles, lóads less díffícúlt.

Carbón Nanótúbe Electrícíty

Carbón nanótúbes are óne óf the new gamers at the renewable electrícíty díscíplíne. These tíny systems are súbmícróscópíc and, as theír call súggests, are fabrícated fróm carbón. The mólecúles línk cóllectívely íntó hóneycómb-shaped systems that shape a túbe that has íncredíble tensíle electrícíty.

Bút ín addítíón tó beíng awesóme stúrdy, recent research have shówn that carbón nanótúbes trúely have the pótentíal tó generate strength. Únder súre ínstances, the nanótúbes can súpply óff pówerfúl waves óf strength.

Íf the researchers at MÍT can hóne the manner, thóse nanótúbes óúght tó pówer small electríc hóme eqúípment wíthín the clóse tó fútúre.

3-D-Prínted Sólar Energy Trees

Íf yóú weren't already satísfíed we líve ín the destíny, yóú have tó take anóther gander at that headlíne: three-D-públíshed sún strength tímber. That's próper, parents; we can líterally prínt trees tó be able tó harness sún electrícíty fór ús. These devíces are prótected ín tíny synthetíc leaves whích míght be made fróm flexíble natúral sún cells.

Each leaf has a separate electrícíty cónverter, and símílarly tó sún energy, thóse tímber alsó can harvest kínetíc and warmth strength íf they're placed óútdóórs. The trúnks óf thóse búshes are crafted fróm wóód-prímaríly based bíócómpósítes, makíng them really líke real tímber.

And nót líke tradítíónal sólar panels, thóse tímber are very appealíng – by way óf far the góód garden órnament ón yóúr blóck.

Hóúse Battery

Ímagíne fór a móment that a resídence, íts persónal strúctúre, was capable óf stóríng pówer. Ín óther wórds, each óf the brícks that make úp íts partítíóns may want tó act as a battery. That's the súdden, thóúgh technícally feasíble, ídea óf a researcher ón the Úníversíty óf Washíngtón. Júlíó D'Arcy and hís gróúp were óperatíng ón the chemístry óf rúst, whích gíves the plaíns óf Mars íts reddísh cólóratíón. And the ídentícal ís góíng fór brícks útílízed ín creatíón dówn ón Earth. Bóth própórtíón hematíte ór ferríc óxíde, the precúrsór míneral tó írón. As súch, hematíte ís cóndúctíve and míght óperate as an electróde. Íf yóú úpload the brícks' póróús mícróstrúctúre, the óppórtúnítíes start tó becóme greater apparent.

Ónce the theóry were ínstalled, D'Arcy and hís crew permeated twó vapórs vía the mícróstrúctúre óf a bríck. When encóúntering the hematíte óre, these generated a pólymer called PEDÓT. Ín thís revólútíónary technólógy task, a bríck wíth 8% hematíte became úsed, whích became a materíal able tó stóríng and releasíng strength thanks tó the remedy. They were ín a pósítíón tó shów ón an LED míld fór fíve mínútes wíth 3 wellknówn brícks.

These new brícks aren't dense súffícíent tó shóp bíg qúantítíes óf strength, as ít ís able tó be óbserved. Hówever, they dó próvíde númeróús advantages. Fór óne aspect, they're tremendóúsly cheap tó pródúce. Alsó, they cóúld face úp tó a cóúple óf chargíng and díschargíng cycles wíthóút dróppíng theír hómes. Ín D'Arcy's test, the brícks have been súbjected tó ten thóúsand cycles whílst preservíng 90% effícíency. And thírdly, they retaín tó featúre nó matter temperatúre ór raín.

Íf we keep ín mínd that a resídence typícally cónsísts óf húndreds óf brícks, ít'd nów nót be únreasónable tó súppóse that theír electrícíty garage capacítíes, partícúlarly fróm sún panels, óffer a actúal benefít. Ín realíty, the develóper óf thóse módern batteríes cónsíders that fífty brícks cóúld be enóúgh tó electrícíty the emergency líghtíng fíxtúres fór fíve hóúrs. Índeed, thís type óf answer wíll cóntríbúte tó móre sústaínable cónstrúctíón.

Sand batteríes

Brícks aren't the handíest nót únúsúal clóth candídate fór the ímpróvement óf módern batteríes. Researchers ón the Úníversíty óf

Calífórnía have exploréd the póssíbílíty óf úsíng sand as a factór fór a versíón that lasts three tímes lónger than the cóntempórary ónes. Óne óf the gróúp cóntríbútórs – recóllect, that ís Calífórnía - drew súggestíón fróm the seashóre sand whílst óút brówsíng. The resúlt ís a cóín-sízed battery that makes úse óf sand fór íts anóde ín place óf cónventíónal graphíte. The fírst step óf the techníqúe túrned íntó tó fínd a sórt óf sand rích ín qúartz, addítíónally knówn as sílícón díóxíde. Ít was then flóór very fínely, ón a nanómeter scale, and then púrífíed tó gaín the qúartz. Fínally, they míxed ít wíth salt and magnesíúm and heated ít. The salt retaíns the warmth, whíle the magnesíúm has the níce óf sóakíng úp the óxygen, só that ín the lóng rún, freed fróm óxygen, the qúartz have becóme sílícón.

The búílders póínt óút that the sílícón óbtaíned ís ín a póróús cóúntry, whích íncreases the úsable flóór place and múltíplíes cóndúctívíty. Whó ís aware óf íf thís kínd óf battery wíll ímpróve the módern-day líthíúm fashíóns' dúrabílíty, assístíng óúr cellúlar telephónes and óther gadgets tó ín the end make ít thróúgh the day.

Órange Ís the New Green Energy

The cóncept that an electríc car, as an example, ís íntrínsícally greener ór extra ecó-fríendly, has íts rísks. What are the assets óf the energy ít cónsúmes? Hów are íts batteríes synthetíc? Óptíng fór the develópment óf renewable energíes ís óne a part óf the eqúatíón; the alternatíve ís that the manúfactúríng methóds óf electrícal cars (EVs) cóme tó be greater sústaínable. And batteríes are at the heart óf thís era. Ríght nów, they make úp a thírd óf the fee óf an electríc aútómóbíle. Fíndíng alternatíves tó cúttíng-edge súbstances ór recyclíng them míght be óf the fúndamental prócesses tó achíevíng ínexperíenced credentíals. Só sóme dístance, hígh-temperatúre treatments have been úsed tó reúse the valúable metals ín batteríes, wíth the drawback óf generatíng tóxíc gases. Hówever, a set óf scíentísts at Nanyang Technólógícal Úníversíty ín Síngapóre ís túrníng tó an answer that úses frúít peels tó reúse the valúable metals ín óld batteríes. Ín thís manner, they can leverage órganíc waste at the same tíme as recyclíng them. Ín dífferent phrases, a manner óf applyíng the cóncepts óf the róúnd ecónómíc system.

Hów are electríc batteríes recycled?

Ín latest years, chemícal treatments have started fór úse ón antíqúe batteríes, whích míght be shreded and decreased tó a paste knówn as black mass. The acíd túb permíts the móst valúable materíals tó be extracted, bút ít's far nevertheless nó lónger a very green answer. The researchers at

Nanyang, then agaín, have úsed óven-dríed and flóór órange peel, tógether wíth cítríc acíd, tó extract cómpóúnds cónsístíng óf manganese, líthíúm, cóbalt, ór níckel wíth an perfórmance óf 90%. Thís ís eqúal tó the resúlts receíved wíth hydrógen peróxíde, óne óf the móst úsúally úsed acíds ín the recyclíng techníqúe.

The researchers póínt óút that óne óf the keys líes wíthín the cellúlóse ín órange peelíngs, that ís cónverted íntó súgars whílst súbjected tó heat thróúghóút the extractíón techníqúe. Ít appears that óther antíóxídants whích inclúdes phenólíc acíds ór flavónóíds, alsó gíft ín the waste óf óranges, addítíónally help tó óptímíze the prócess. Óne óf the blessíngs óf the brand new appróach, except íts lów charge, ís that the resúltíng resídúes are nón-tóxíc.

A technólógy ventúre wíth (real) búsíness prógrams

Freqúently, thís sórt óf technólógícal challenge wórks ín a labóratóry envírónment bút then faíls tó transítíón íntó búsíness manúfactúríng. Hówever, develópers have already created púrpósefúl batteríes fróm recycled súbstances that shów a símílar chargíng abílíty tó the aúthentíc devíces. They are nów engaged ín methód óptímízatíón tó ímpróve the perfórmance óf recycled batteríes símílarly and óptímíze theír manúfactúríng ón a massíve scale. Technícally, órange peelíngs are óne óf the óptíóns, bút ít ís próbably feasíble tó apply óther vegetable waste, whích they're already readíng. Alsó, they're explóríng the óppórtúníty óf applyíng thís módern appróach tó batteríes óf díverse kínds, whích inclúdes líthíúm, írón, and phósphate.

Shadóws Renewable Energy

Ín the fíeld óf renewable energíes, there ís exístence past phótóvóltaíc ór wínd strength. Fór example, we've póínted óút óptíóns inclúsíve óf wave strength ór we've even inclúded úsíng sweat as a súpply óf pówer fór wearables. There ís even the póssíbílíty óf manúfactúríng energy fróm the snów. These generatíón tasks shów that strength ís anywhere and that ít best takes a tóúch íngenúíty and an awesóme dóse óf engíneeríng tó harness ít. Anóther próóf óf thís ís the revólútíónary tóól created vía researchers ón the Úníversíty óf Síngapóre. Íf there's sóme thíng greater cónsíderable than líght, ít ís the shadów, and that resóúrce ís what they've taken gaín óf tó generate small electríc cúrrents. They have named ít SEG (shadów ímpact strength generatór), and ít may strength many famíly devíces ín a nót tóó remóte fútúre.

Researchers factór óút that latest phótóvóltaíc cells want a nón-stóp súpply óf míld and that the ínterrúptíón óf thís ínflúences theír pówer performance. Tó allevíate thís tróúble, thís crew taken íntó cónsíderatíón the póssíbílíty óf takíng gaín óf the líghtíng fíxtúres cóntrasts pródúced by means óf shadóws as an índírect electrícíty súpply. They declare that thís ís an remarkable methód and wíth remarkable pótentíal fór grówíng devíces whích can paíntíngs each índóórs and óútsíde, whereín the próvísíón óf líght ís úsúally extra díscóntínúóús. The era they have advanced ís each lów príced and self-pówered.

Nanótextúre ínfógraphíc

Hów dóes the shadów ímpact paíntíngs?

The ínítíal devíce they've created cónsísts óf a seríes óf SEG cells alígned ón a óbvíóús plastíc membrane. Each óf these cells has twó layers: a sílícón súbstrate and a thín góld fílm. When the cómplete flóór óf the devíce ís úncóvered tó míld, the present day óf strength ít generates ís very súsceptíble. And the ídentícal ís actúal whíle ít's far cómpletely shaded. Hówever, whíle the móbíle ís handíest partíally íllúmínated, a extensíve electríc present day ís pródúced, becaúse the cell túrns íntó each a generatór and a cóllectór óf electrícíty. The fírst labóratóry checks shów that the same óld líghtíng cóndítíóns ín a hóme make ít feasíble tó generate an electríc cúttíng-edge óf 1.2 V, í.E., stróng enóúgh tó rún a dígítal watch.

Ín addítíón tó íts úse ín wearables and smartwatches, researchers agree wíth that ít can be úsed ín self-pówered sensórs. Thús, whenever a mótíón óccúrs that alters the ambíent líghtíng, the sensór may be actívated róútínely. Anóther vícíníty óf ínterest wóúld be embeddíng these PV cells íntó clever garb. Fínally, phótóvóltaíc panels may be created tó be úsed wíthín the hóme.

The develópers óf the SEG cónsíder that theír manúfactúríng cóst cóúld be lówer than that óf cónventíónal sílícón cells. Tó thís end, they're cónsíderíng changíng góld wíth óther, extra lówer príced materíals. Ín bríef, tó take every óther step tówards the úse óf greater sústaínable technólógy.

ᗺᗺᗺ

THREE

CHAPTER-2

New Generatíón óf Wind Túrbínes Energy

The úpward púsh óf renewable pówer maíntaíns únstóppable. Alóng with new technólógíes at the ríse, ínclúsíve óf wave electrícíty, the ónes already cónsólídated are beíng perfected and stepped fórward óf theír performance. Ín the case óf wind strength, óne óf the latest advances has been tó cóúple ít wíth sólar energy. The techníqúe inclúdes óverlayíng the wínd túrbíne tówer wíth phótóvóltaíc sún panels capable óf generatíng pówer tó delíver the ínner strúctúres óf the túrbíne. Óften, whílst wind generatórs remaín ídle becaúse óf lack óf wind, they reqúíre tó keep sóme manage systems wórkíng. Úsúally, they may be related tó the electríc grid tó delíver theír desíres, hówever nów the sún panels wíll take óver the ventúre. The pílót check has been carríed óút ínsíde the Breña wind farm (Albacete), ín óne óf the wind mílls that ACCÍÓNA ówns.

The setúp capabílítíes óne húndred twenty sólar panels that óccúpy fífty meters óf peak óf the tówer, wíth a energy óf 9.36 kílówatts peak. The panels have been íncórpóráted wíth a sóútheast-sóúthwest óríentatíón tó óptímíze the captúre óf sólar electrícíty. The íntegratíón óf the PV sún panels ísn't always the ónly revólútíónary element óf the task, wíth the chóíce óf a brand new generatíón óf natúral sólar panels. Barely óne míllímeter thíck, they're characterízed by way óf úsíng carbón as raw fabríc, próvíde bríllíant strúctúral flexíbílíty and líghtness, and are entírely recyclable. They addítíónally have lówer maíntenance charges. Althóúgh fór nów theír perfórmance ís póórer than sílícón-prímaríly based módels, thís new fórm óf panel has extremely góód capacíty, becaúse the task's prómóters póínt óút.

"Breña's hybrídízatíón míssíón inclúdes óptímízíng the úsage óf area fór renewable energy manúfactúríng and ís góíng tó allów ús tó check the perfórmance óf órganíc phótóvóltaícs, a generatíón that we accept as trúe wíth has óne óf the maxímúm technícal perfórmance ímpróvement cúrves. That ís why we've determíned tó check ít", saíd Belén Línares, Díректór óf Energy Ínnóvatíón at ACCÍÓNA.

Órganíc sólar panels fór hóme úse

Althóúgh they wíll nów nót replace sílícón sólar panels ín indústríal packages, fór nów, órganíc panels may want tó have thríllíng hóme applícatíóns, maínly fór ÍóT technólógíes. An ínstance óf thís ís the latest annóúncement by way óf the French Cómmíssíón fór Alternatíve Energy and Atómíc Energy. The French frame has annóúnced the effects óf stúdíes ín cóllabóratíón wíth the Japanese córpóratíón Tóyóba: an natúral panel wíth a cónversíón fee óf 25% ín a dark róóm wíth 220 lúx neón líghtíng fístúres. Thís perfórmance ís 60 % hígher than that óf amórphóús sílícón phótóvóltaíc cells, whích míght be óften úsed ín pócket sún calcúlatórs.

The Japanese agency has advanced a sólúble materíal able tó generatíng electrícíty by means óf órganíc synthesís era. Thús, ít ís able tó be díssólved very easíly and carríed óut flíppantly tó a súbstrate. These new phótóvóltaíc cells can wórk where tradítíónal cells cannót and can accórdíngly strength dómestíc gadgets ín póórly lít róóms índependently óf the electrícíty grid.

Wí-Fí Netwórks Energy

Óne óf the maxímúm freqúent rítúals úpón arrívíng at a new hóme ís ínqúíríng fór the Wí-Fí passwórd. These úbíqúítóús netwórks, avaílable at hómes, mótels, aírpórts, cóffee shóps, and óther públíc spaces, próvíde speedy and, freqúently, lóóse cónnectívíty. Hówever, they míght qúíckly be súpplyíng a cómpletely úníqúe featúre — úsíng terahertz waves, knówn as T-rays, díscóvered ín Wí-Fí alerts as an revólútíónary súpply óf pówer. Devíces líke smartphónes ór smartwatches óught tó as a cónseqúence be recharged ín any vícíníty ínsíde the range óf thís electrómagnetíc radíatíón. Thís sórt óf electrícíty became só far únúsable. Stíll, researchers at MÍT agree wíth that Wí-Fí netwórks shóúld end úp a brand new way óf transmíttíng electrícíty wírelessly.

Graphene óver agaín

The team, led vía Híróki Ísóbe, a member óf MÍT's Materíals Research Lab, cúrrently pósted a píece óf wrítíng wíthín the jóúrnal Science Advances próvíng the feasíbílíty óf thís era. And nót símply ón paper, as the researchers are already rúnníng ón a bódíly devíce. Their methód ís based tótally ón the úsage óf graphene and íts cóndúct ón a qúantúm scale. The crew has verífíed that, by cómbíníng graphene wíth óther materíals líke bórón nítríde, the electrón flúx can be managed and skewed clóser tó a únmarríed dírectíon. Thús, the fabríc wóúld deal wíth terahertz waves as a vísítórs warden, channelíng them thróúgh a únmarríed lane and rewórkíng them íntó a díred módern (DC).

Prevíóús tests had cónverted lów-freqúency radíó waves íntó a ríght away módern hówever have been nót able tó reap terahertz waves, that cóúld generate a strónger módern. Ónly the úsage óf últracóld temperatúres had presented a súccess resúlts. Únfórtúnately, thís kínd óf set úp preclúded móst practícal applícatíons. The only óppórtúníty changed íntó tó úse a clean clóth, inclúding graphene, tó góvern the díred módern at róóm temperatúre.

Wórkíng ón that precept, Ísóbe has recómmended a small graphene sqúare wíth a layer óf bórón nítríde and an antenna. Graphene ís a symmetríc clóth, whích means that electróns receíve íncómíng electrícíty waves fróm all gúídelínes and scatter ín all gúídelínes tóó. The úse óf bórón nítríde, bút, alters the symmetry, as bórón attracts electróns ín a síngle path and nítrógen ín a dístínctíve óne. Thís tensíon steers electróns ín a únmarríed róúte, fór that reasón prodúcíng a díred cúttíng-edge. Researchers examíne thís technólógy tó a sólar PV móbíle that captúres electrómagnetíc waves ín place óf daylíght.

The last energy súpply fór the Ínternet óf Thíngs?

Ín the fóllówíng few years, a número óf recent Ínternet-enabled mícródevíces wíll arríve ínsíde the market. All kínds óf sensórs, embedded ín a húge varíety óf óbjects, wíll talk amóng themselves and wíth the cómmúníty wíthóút the want fór an Ínternet brówser. The Ínternet óf Thíngs (ÍóT) ís póised tó emerge as a gamechanger, hówever all the ónes gadgets wíll want their pówer assets. Technícally, T-rays cóúld be a way ó addressíng these reqúírements. Devíces súch as wearables, pacemakers, ór óther frame ímplants cóúld alsó gaín fróm thís wíreless chargíng generatíon.

Blúe Energy

Íf yóú want a bríef remínder óf what blúe pówer ís ín realíty abóút, yóú may test thís vídeó. Júst tó súm úp, blúe pówer ór salíníty gradíent electrícíty harnesses the dífferentíal amóng freshwater and saltwater, as ít óccúrs ín a ríver flówíng íntó the ócean, tó generate pówer. Althóúgh thís sóúrce óf electrícíty has been recógnísed fór a few a lóng tíme by way óf nów, ít ís stíll ín íts early ranges óf ímplementatíón. Ín trúth, óne óf the fírst blúe strength plants was ópened ín Nórway as lately as 2009. Prómísíng prógress, hówever, has been made ón thís prógressíve technólógy. A Stanfórd Úníversíty labóratóry has created a new battery prímaríly based ón blúe energy. They have referred tó as ít Entrópy Míxed Battery (EMB).

The blúe pówer devíce develóped vía the Stanfórd engíneers dístíngúíshes ítself fróm prevíóús technólógíes by way óf the trúth that ít dóes nów nót reqúíre stress (called straín retarded ósmósís ór PRÓ) ór membranes (óppósíte electródíalysís ór RED), generatíng pówer cómpletely thróúgh an electróchemícal techníqúe. The battery carríes a tank thís ís crammed úp wíth efflúent fróm A wastewater treatment plant. Several electródes ínsíde the tank release sódíúm íóns whíle súbmerged ín the water. The mótíón óf these íóns creates an electríc módern flówíng fróm the anóde tó the cathóde. Next, freshwater ís únexpectedly replaced wíth saltwater, whích sends the sódíúm and chlóríde íóns lówer back tó the electródes, ínvertíng the strength módern. Bóth the freshwater and saltwater ínflúxes generate pówer, whích means that the battery ís charged and díscharged cóntínúóúsly wíth óút the need óf external strength sóúrces.

"The fírst checks wíth the blúe pówer battery, prímaríly based ón the alternate óf saltwater and seawater, had been fíníshed ín a wastewater treatment plant ín Paló Altó (Calífórnía)."

The fírst checks wíth thís blúe strength technólógy ventúre fíníshed ín a wastewater treatment plant ín Paló Altó (ÚSA), swítchíng fróm saltwater óbtaíned fróm the clóse by bay tó already handled freshwater ín óne-hóúr cycles, have próved the feasíbílíty óf thís new technólógy. The researchers have shówed that, thróúghóút óne húndred eíghty cycles, the materíals have maíntaíned a nínety seven percent perfórmance even as takíng píctúres the salíníty gradíent energy.

The EMB ís the secónd óne generatíón óf thís sórt óf batteríes that they have evólved. The fírst módel úsed steeply-príced sílver-prímaríly based electródes wíth cónstraíned búsíness applícatíóns. Nów, rather than sílver, the electródes are cóvered wíth Persían blúe, a very lów-valúe pígment, whích sells fór múch less than a greenback accórdíng tó kíló, cóllectívely wíth pólypyrróle, than can be óffered fór 3 dóllars a kíló ín búlk.

A bíg abílíty

The púrpóse fór EMB ís tó óffer wastewater remedy vegetatíón wíth a súpply óf strength that makes them carbón ímpartíal and pówer self-enóúgh. The essentíal díffícúlty wíth thís kínd óf blúe electrícíty vegetatíón ís that they're presently strength-íntensíve and that they're úncóvered tó electrícíty cúts whích can bóg dówn theír óperatíóns. Íf the freshwater laúnched íntó the ócean became a súpply óf electrícíty tó electrícíty thóse plant lífe, the wastewater remedy cycle shóúld becóme extra sústaínable and ecó-fríendlíer.

As ít's been talked abóút, cóastal wastewater remedy plant lífe are cónsídered óne óf the bíggest póssíbílítíes tó harness thís kínd óf blúe pówer. Sóme estímates endórse that, íf all the flówers óf thís type ínsíde the ínternatíónal were úsed líke thís, they may generate appróxímately 18 gígawatts.

Alternatíve Energy Sóúrces óf the Fútúre

Despíte the hype acróss the prógress óf renewable electrícíty, many peóple dón't realíze that sólar and wínd have móst effectíve made a tíny dent wíthín the electrícíty míx úp tó nów. The córrect news ís that cósts are cómíng

dówn and many húman beíngs are startíng tó úndertake ínexperíenced technólógy, bút there may be nevertheless a móúntaín tó clímb íf we want tó síncerely get óff óf fóssíl fúels ón a húge scale.

Tó accómplísh thís, we're góíng tó óught tó thínk óútdóór the bóx tó próvíde yóú wíth new appróaches tó address the pówer próject. Lúckíly, the parents at Fútúrísm have pósítíóned ten óf the maxímúm prómísíng alternatíve pówer resóúrces óf the destíny ín a handy ínfógraphíc. Sóme óf these may be lengthy píctúres, bút a few may alsó play a essentíal róle ínsíde the strength míx óf the destíny.

Space-prímaríly based sólar

Móst sólar pówer dóesn't genúínely make ít íntó the Earth's envírónment, só area-based sún energy makes a lót óf feel. The challenges are the cóst ín gettíng a satellíte tv fór pc tó órbít, as well as the cónversíón óf pówer íntó mícrówaves that may be beamed ríght dówn tó the planet's flóór.

Húman electrícíty

There's óver seven bíllíón húmans stróllíng acróss the Earth every day, só why nó lónger generate energy fróm the mótíón óf húman beíngs? Many prófessíónals belíeve that we can harness thís electrícíty, and that we shóúld úse ít tó energy óúr gadgets.

Tídal energy

Fíve ínternatíónal lócatíóns aróúnd the sectór are begínníng tó perfórm póssíble wave electrícíty farm óperatíóns, bút the pótentíal ís sóme dístance hígher: the Ú.S. Cóastlíne by myself has a wave energy capacíty óf abóút 252 bíllíón KWh ín step wíth 12 mónths.

Hydrógen pówer

Hydrógen ís a easy and róbúst súpply óf strength, and níce óf all – ít debts fór 74% óf the mass óf the whóle úníverse. The móst effectíve próblem ís that hydrógen atóms have a tendency tó símplest be lócated ín cómbínatíóns wíth óxygen, carbón, and nítrógen atóms. Remóvíng thís bónd takes energy, whích ends úp beíng cóúnter-pródúctíve. As a resúlt, many húman beíngs aróúnd the sectór are óperatíng ón makíng these appróaches greater fínancíal.

Magma electrícíty

The míddle óf the Earth cóúld be very warm, só why nót try tó get clóser tó ít tó tap íntó a few geóthermal warmth? Peóple ín Íceland are already dóíng thís wíth crímsón-hót magma after accídentally placíng a pócket óf ít ín the cóúrse óf a 2008 dríllíng task.

Núclear waste

Ónly fíve% óf úraníúm atóms are útílízed ín a cónventíónal físsíón respónse. The rest túrn óút tó be ínsíde the píle óf núclear waste, whích síts ín garage fór lóts óf years. Researchers and agencíes are tryíng tó tap íntó these leftóvers fór a feasíble and fínancíal energy sólútíón.

Embeddable sún strength

What íf every wíndów míght be easíly was a sólar panel? Sólar wíndów technólógy túrns any wíndów ór sheet óf glass ríght íntó a phótóvóltaíc sún cellúlar that harvests the part óf the líght spectrúm that eyes can't see.

Algae strength

Algae gróws nearly everywhere, and ít seems thóse tíny plants are a súrprísíng súpply óf electrícíty-rích óíls. Úp tó 9,000 gallóns óf bíófúel may be "grówn" per acre, makíng ít cónsídered óne óf many abílíty energy sóúrces óf the destíny.

Flyíng wínd electrícíty

Wínds are a great deal móre pówerfúl and stúrdy at better elevatíóns. Íf wínd farms may be self súffícíent and flyíng, they may gó tó ín whích the wínds are móst pówerfúl and delíver dóúble the pówer óf símílarly sízed tówer-estáblíshed túrbínes.

Fúsíón electrícíty

Fúsíón has been the dream fór sóme tíme – bút scíentísts are makíng tóddler steps tó attaíníng the strength prócedúre that ís harnessed ín natúre by means óf óúr very ówn sólar. The ÍTER (Ínternatíónal Thermónúclear Experímental Reactór) ís presently beíng búílt ín France, and ít's óne óf the maxímúm cómplícated clínícal and engíneeríng íníatíves ín exístence.

ᗪᗪᗪ

FOUR

CHAPTER-3

The Fútúre óf Pówer

The húman race, ín íts never ending war tó impróve íts preferred óf resíding, has always relíed ón large amóunts óf electrícal electrícíty tó gas óur evólútíon. A cúttíng-edge estímate by way óf Natíónal Geógraphíc decíded that we úse 320 bíllíon kílówatt-hóurs óf strength each day. Tóday, maxímúm óf thís extensíve reqúírement ís addressed vía búrning fóssíl fúels. Só sóme dístance, fóssíl fúels have catered tó óur pówer desíres very córrectly, bút they're alsó nón-renewable and swíftly depletíng. These gasólíne resóurces have addítíónally cóntríbúted greatly tó greenhóuse gas emíssíóns and póllútíón. The tíme has cóme tó díscóver apprópríate and hígher replacements fór fóssíl fúels. Scíentísts are cóntínúóusly stúdying móre recent and greener assets óf energy whích have restrícted impact ón the envírónment and decrease theír cóntríbútíon tó glóbal warmíng, whích ís thóught tó be resúltíng fróm the release óf carbón díóxíde at the same tíme as búrning fóssíl fúels.

Atómíc electrícíty, sún pówer, and electrícíty fróm wínd and bíó fúels are ónly sóme óf the prómísíng óptíóns fór a púrífíer and greener fútúre. Óther fantastícally new resóurces óf electrícíty tógether wíth gasólíne cells, geóthermal pówer, and ócean electrícíty alsó are beíng explóred. Ín the súbseqúent sectíóns, we'll test módern resóurces óf pówer as well as talk feasíble fútúre pówer resóurces.

Fóssíl Fúels – CóalCóal Pówered Pówer Plants (Fóssíl Fúel)

Fóssíl fúels are the remaíns óf dead flóra and anímals ón land and ín the seabed. These are fashíoned fróm the fóssílízed stays óf lífeless anímals and plant lífe whích are úncóvered tó heat and stress ínsíde the earth's crúst fór lóads óf húndreds óf thóúsands óf years.

Fóssíl fúels freqúently encómpass hydrócarbóns. They íncórpórate carbón and hydrógen ín varíóús ratíós, tógether wíth methane, that has a lów carbón tó hydrógen ratíó, ór anthracíte cóal, whích ís nearly púre carbón. Hydrócarbóns are fashíoned whíle the fóssílízed stays óf úseless órganísms are chemícally altered óver lóads óf thóúsands and thóúsands óf years by excessíve stress and warmth díscóvered wíthín the earth's crúst. The chemícal energy 'stóred' ín these fúels ís released at sóme póínt óf cómbústíon tó súpply electríc energy.

Accórdíng tó estímates súpplíed vía the Energy Ínfórmatíón Admínístratíón, fóssíl fúels accóúnt fór 86% óf the fúll energy pródúced wíthín the glóbal. Óf thís, petróleúm accóúnted fór 36.Eíght%, cóal 26.6% and natúral gas 22.Níne%.

Hówever, fóssíl fúels are nón-renewable sóúrces óf strength. They take lóads óf húndreds óf thóúsands óf years tó shape and are depleted a lót faster than new reserves can be created. Ít ís expected that 23.5 tóns óf fóssílízed órganíc clóth depósíted ón the sea gróúnd ís needed tó pródúce 1 líter óf fúel. Ín 1997, the tótal amóúnt óf fóssíl gasólíne úsed túrned íntó eqúívalent tó plant remember that grew ón the cómplete land and ócean súrface óf the earth óver a dúratíón óf 422 years.

Anóther drawback óf óúr heavy dependence ón fóssíl fúels ís the qúantíty óf carbón díóxíde pródúced at sóme stage ín cómbústíon, whích ís estímated at 21.Three bíllíón heaps ín keepíng wíth year. Hówever, natúral prócedúres are capable óf sóakíng úp ónly appróxímately half óf óf the óverall amóúnt óf carbón díóxíde emíssíóns laúnched íntó the ecósystem, whích means that each year the qúantíty óf carbón díóxíde wíthín the atmósphere ís íncreasíng by means óf 10.Síxty fíve bíllíón heaps, that ís theórízed tó be the leadíng cóntríbútór tó ínternatíónal warmíng that wóúld dóúbtlessly have very únfavóúrable óútcómes at the atmósphere.

Fóssíl Fúels – Natúral Gas

Natúral gasólíne ís typícally determíned alóng wíth fóssíl fúels, ín cóal-beds and trapped ín dífferent sórts óf róck. Ít ís created by way óf methanógeníc órganísms present ín landfílls, marshes and wetlands. Ít evídently cónsísts óf methane and small amóúnts óf dífferent gases súch as ethane, própane, bútane, pentane, hydrócarbóns óf hígher mólecúlar weíght, súlfúr, helíúm and nítrógen. The cómpónents óf natúral fúel apart fróm methane need tó be remóved earlíer than natúral gas can be úsed as a sóúrce óf gas. Read Natúral Gas Generatórs: An Alternatíve tó Díesel, fór óne example shówíng present era úsíng a natúral aíd, óne thís ís hígher fór the súrróúndíngs, as gas.

Althóúgh herbal gas ís cónsídered tó be cleaner than óther fóssíl fúels, ít has stíll been fóúnd tó cóntríbúte tó póllútants and ínternatíónal warmíng. Whíle ít cóúld be úsed tó súpplement the wórld's ever depletíng reserves óf tradítíónal fóssíl fúels, ít ísn't a 100% smóóth, nón-póllútíng alternatíve. Ín 2004, carbón díóxíde emíssíóns dúe tó úsíng natúral gasólíne stóód at 5,three húndred míllíón heaps whílst cóal and óíl cóntríbúted tó carbón díóxíde emíssíóns óf 10,síx húndred míllíón tóns and 10,twó húndred míllíón tóns, respectívely. Hówever, thís fashíón ís expected tó reverse vía 2030 when herbal gasólíne ís póssíbly tó emít 11,000 míllíón heaps óf carbón díóxíde ín place óf eíght,400 míllíón lóts fróm cóal and 17,2 húndred tóns fróm óíl at that póint. Alsó, whílst released ímmedíately íntó the atmósphere, natúral gasólíne ís a múch móre pótent greenhóúse fúel than carbón díóxíde hówever gíven that thís óccúrs ín very small amóúnts, ít's far cúrrently nót a maín púrpóse óf sítúatíón.

Sólar Energy

Almóst the entírety ón thís ínternatíónal ín the lóng rún deríves íts electrícíty fróm the sólar. Ínstead óf acqúíríng the sún's strength fróm óblíqúe assets líke fóssíl fúels, researchers and agencíes glóbal are tryíng tó ímmedíately tap thís límítless súpply óf electrícíty.

The earth receíves abóút 174 bíllíón megawatts óf strength at the hígher envírónment becaúse óf sólar radíatíón. Abóút 30% óf the íncídent sún radíatíón ís cóntemplated back, at the same tíme as the fínal, whích amóúnts tó 3.85 x 1024 Jóúles every year, ís absórbed thróúgh the ecósystem, óceans and landmasses. The qúantíty óf sún strength thís ís tó be had tó ús at sóme stage ín an hóúr ís extra than the fúll amóúnt óf pówer ate úp wórldwíde ín a whóle 12 mónths. Bút thís ís a díffúsed, rather than fócúsed, fórm óf strength and the greatest task líes ín harnessíng ít.

Heat and míld radíatíón fróm the sún can be harnessed thrú úsíng semícóndúctór sólar panels. The strength sólar radíatíón excítes electróns ón thóse panels and ends ín the manúfactúríng óf electrícal pówer.

Óne óf the móst ímpórtant húrdles ín harnessíng the electrícíty fróm the sún ís ín búíldíng príce-effectíve sún panels. The valúe óf sún pówer ís abóút ÚS 8–15 cents ín líne wíth kílówatt-hóúr ín cómparísón tó the cóst óf cóal-based electríc pówered pówer at ÚS 6 cents ín líne wíth kílówatt-hóúr.

Próper stórage óf pówer ís every óther majór óbstacle. Sólar electrícíty ísn't aváilable at níght tíme hówever módern-day electrícíty systems úsúally antícípate nón-stóp aváilabílíty óf electrícíty. Thermal mass systems, thermal garage strúctúres, segment trade súbstances, óff-gríd phótóvóltaíc systems, and púmped stórage hydróelectrícíty systems are a númber óf the appróaches whereín sólar strength can be stóred fór later úse.

Even wíth all óf the technólógícal advancements, sún pówer technólógy ís stíll ín íts ínfancy. Úntíl we perfect the generatíón and are able tó harness and keep sún pówer ín a feasíble and valúe-pówerfúl way, fóssíl fúels wíll cóntínúe tó be the maxímúm cómmónly úsed sóúrce óf energy.

Núclear Energy

Núclear Pówer Plant

As the wórldwíde demand fór electrícíty maíntaíns tó súrge, núclear energy ís gaíníng íncreasíng sígnífícance as a clean súpply óf energy thís ís predícted tó address the glóbal próblem óf clímate trade. Vólatílíty wíthín the príces óf fóssíl fúels and the grówíng próblem óf cóúntríes tó relaxed energy súbstances are óther drívers óf núclear electrícíty.

There are cúrrently 439 núclear electrícíty reactórs óperatíónal ín 30 ínternatíónal lócatíóns ínternatíónal. Thís accóúnts fór 14% óf the tótal strength generatíón óf the sectór. The Ínternatíónal Atómíc Energy Agency (ÍAEA) expects the wórldwíde núclear electrícíty generatíón capacíty tó bóóm fróm the cúrrent 372 gígawatts (GW) tó 437–542 GW by way óf 2020 and tó 473–748 GW thróúgh 2030. Hówever, fór núclear electrícíty tó becóme a dependable and smóóth súpply óf strength, several challenges want tó be addressed. Sóme óf these ínclúde ímpróvement ín mónetary cómpetítíveness, desígníng safe and relíable núclear electrícíty plant lífe, management óf spent gas and díspósal óf radíóactíve waste, grówíng ók skílled gróúp óf wórkers, ensúríng públíc self belíef ín núclear electrícíty, and makíng súre núclear nón-prólíferatíón and prótectíón.

Núclear electrícíty ís harnessed by way óf eíther splíttíng (físsíón) ór mergíng (fúsíón) the núcleí óf twó ór extra atóms. Núclear físsíón cómmónly úses úraníúm ín the manner óf harnessíng strength. At óúr módern fees óf cónsúmptíón, the úraníúm determíned ín the Earth's crúst can remaíníng ús appróxímately a centúry. Hówever researchers expect that the pówer íntake wíll tríple ín the next centúry, whích means that the aváílable úraníúm sóúrces wíll handíest last ús fór abóút 30 years. Óne alternatíve ís the reprócessíng óf the spent gas. Thís spent gas ís wealthy ín plútóníúm and whilst blended with the leftóver úraníum, ít may be reprócessed ríght íntó a aggregate called MÓX, which can be úsed as gas. Thís may help tó stretch the tó be had úraníum sóúrces by way óf a few móre decades. The largest dísadvantage tó thís súpply óf strength ís the dispósal óf radióactíve waste and the excessíve fee óf búíldíng núclear energy plant life.

Núclear físsíón, hówever, míght be the sólútíón tó óúr electrícíty próblems. Físsíón útílízes hydrógen ísótópes, líthíúm, and bórón. The líthíúm reserves fróm the earth, míxed wíth thóse fróm the ócean, can clósíng ús fór greater than 60 míllíón years. Deúteríúm, an ísótópe óf hydrógen, can clósíng any óther 250 míllíón years. Hówever, the techníqúe óf harnessíng pówer fróm thís ísótópe ís faírly cómplícated and cóntínúes tó be ín íts ínfancy. Íf we can effectívely learn hów tó útílíze núclear fúsíón fór the era óf electrícíty ín a feasíble way, ít can próperly be the new kíng óf the pówer glóbal. Núclear fúsíón ís a easy techníqúe, wíth lów carbón díóxíde emíssíóns, and the radióactíve waste merchandíse actúally have a enórmóúsly shórt half-life.

Wind Energy

Wínd farms are cónstrúcted tó harness mechanícal energy fróm the wínd and cónvert ít íntó electríc electrícíty. These wínd farms are then related tó electrícal electrícíty transmíssíón netwórks fór the dístríbútíón óf electrícíty. Ón cómmón, handíest 20 tó 40 percent óf the whóle energy capabílíty óf a wínd farm can be applíed.

The próscríbíng factór ín harnessíng strength fróm wínd ís that wínd pace ís varíable and ín móst cases the electrícíty fróm wínd can handíest be súccessfúlly harnessed wíth very hígh wínd speed and regúlar heavy wínds. These generally óccúr at hígher altítúdes. Wínd strength alsó calls fór bíg, ópen expanses óf land fór yóú tó assemble wínd farms.

Ín 2008, the wórldwíde wínd strength generatíón pótentíal stóód at 121.2 GW. Ón a medían, wínd energy presently debts fór símplest 1.5% óf the glóbal electrícíty era capacíty. Hówever, thís zóne has grówn -fóld ínsíde the three-yr dúratíón óf 2005–2008. Wínd electrícíty debts fór 19% óf the tótal electrícíty technólógy ín Denmark, 10% ín Pórtúgal and Spaín, and 7% ín the Repúblíc óf Íreland and Germany.

Bíófúels and Bíómass

These cónsíst óf fúel fróm plant and anímal sóúrces. Óíl, ór ethanól, receíved fróm plants cónsístíng óf súgarcane, swítchgrass, algae, póplar, and córn can be úsed ímmedíately ór blended wíth óther fúels inclúsíve óf índústríal díesel and gas tó próvíde electrícíty. Even plant remember tógether with dead wóóden, leaves, tímber chíps, and branches can be búrnt tó próvíde energy. Thís ís generally categórísed as bíómass. Bíómass addítíónally cónsísts óf any bíódegradable waste fróm plant and anímal resóúrces whích can be búrnt fór fúel.

The restrícting factór ín úsíng bíó fúels ís that a large númber óf plants want tó be grówn tó reap the energy trapped ín plant lífe. Thís reqúíres góód sízed regíóns óf fertíle land. Addítíónally, nów nót all plant sóúrces óffer a excessíve yíeld. Experíments are únderway tó hybrídíze and genetícally adjúst these vegetatíón tó make them móre stúrdy and íncrease theír yíeld. Bíófúels are very prómísíng fór small-scale úse as they are lów ón greenhóúse fúel emíssíón, are an pówerfúl waste management devíce, and pródúce líttle aír póllútíón.

With the advancement óf new era and the ímpróvement óf recent ínsíghts íntó óúr envírónment, scíentísts have been able tó cóme úp wíth even móre adventúróús strength óptíóns. These ínclúde fúel cells, geóthermal strength, and tídal and wave pówer, tó call sóme.

Fúel Cells

Fúel cells are símílar tó batteríes hówever úse reactants fróm an external sóúrce, ínstead óf batteríes whích can be self cóntaíned. Íf the gas and óxídant ranges ín gasólíne cells are well maíntaíned, strength may be generated nearly cóntínúóúsly. The efficiency óf fúel cells ís própórtíónal tó the pówer beíng drawn fróm ít. They are alsó líght-weíght and extremely relíable.

Geóthermal Energy

The ínteríór óf the Earth cónsísts óf lóts óf warmth. Shallów areas cómpríse warm water, róck and steam. Deeper ínner, the magma ís very hót. Thís heat may be harnessed tó próvíde electrícal strength and fórce númeróús packages. Harnessíng geóthermal electrícíty calls fór nó fúel and mínímúm land. Ít ís partícúlarly cheap and a tótally sústaínable súpply óf energy cónsíderíng the qúantíty óf heat cóntaíned ín the earth bed ís só vast that even thóúgh we harness móre strength than we reqúíre, ít's góíng tó nónetheless súffíce fór húndreds óf thóúsands óf years yet tó cóme.

Óceaníc Energy

Pówer Generated by Waves

The óceans are húge and cómpríse bíg amóúnts óf energy wíthín the water cúrrents, and thermal and salíníty gradíents. The electrícíty fróm tídes and waves may be harnessed tó próvíde electrícal strength. The dífferences ín temperatúre that aríse wíth varyíng depths may be úsed tó pressúre heat engínes, whích ín túrn pródúce electríc strength.

The ósmótíc pressúre dístínctíón amóng salt water and sparklíng water alsó can be úsed tó generate pówer. Althóúgh móst óf these strategíes are nónetheless wíthín the experímental ranges, íf researched nícely, they can be a step fórward fór mankínd. The óceans can be able tó qúench óúr thírst fór energy and bag the crówn as the kíng óf fúels.

Energy fróm Antímatter:

Óne óf the móst cómplícated theóríes óf manúfactúríng strength ís the ídea óf the úse óf rely and antí-remember tó generate electríc energy. Antímatter ís the óppósíte óf remember. Íf remember ís cónstítúted óf debrís, antí be cóúnted ís created fróm antí-partícles.

Cíentísts advíse that íf matter and antí rely have been tó cóllíde, they wóúld anníhílate each óther and release extensíve qúantítíes óf electrícíty. Hówever, that ís nevertheless a theóretícal sóúrce óf strength. Whether antí-rely exísts ín sóme part óf the úníverse and can be harnessed ín a few way ís stíll a mystery tó húmankínd.

There are díverse methóds óf extractíng strength fróm the earth that húmankínd has determíned and úsed tó íts benefít. As the húman race evólves, we are able tó cónstantly lóók fór newer, extra green varíetíes óf energy whích have the least amóúnt óf ímpact ón the súrróúndíngs. At módern, the maxímúm ecónómícally effícíent fúel has próved tó be óíl.

Ín the fútúre, whíle the sectór's óíl reserves are depleted, we can úse sóme óther sóúrce óf energy; próbably óne thís ís cíted abóve. Hówever, the trúth óf the próblem ís that we shóúld be próactíve ín researchíng new sórts óf strength tó maíntaín the develópment óf cívílízatíón and tó make certaín a hígh hígh-qúalíty óf resídíng that we all have grówn acqúaínted wíth.

Space technólógy

Space-prímaríly based electrícíty technólógíes—thíngs líke harvestíng hydrógen fróm the móón tó electrícíty gas cells ón Earth, ór órbítíng sún arrays that absórb aróúnd-the-clóck díréct daylíght and beam the energy back dówn tó statíóns ón the flóór vía radíó ór mícrówaves—stay fírmly ínsíde the realm óf technólógícal knów-hów fíctíón fór nów.

Bóth NASA and the Ú.S. Naval Research Lab are already makíng an ínvestment ín the technólógy that wóúld be cómmercíalízed ín 25 years. The óngóíng nón-públíc space renaíssance that has seen órganízatíóns líke SpaceX trím the príce óf laúnchíng cargó íntó órbít bódes well fór greater ambítíóús tasks ín space.

Sólaren, a sóuthern Califórnía-prímaríly based begín-úp, has ínked a deal tó delíver Pacífíc Gas and Electríc wíth space-based sólar strength vía the cease óf the last decade.

Fúsíón: The Fíre Sóme Tíme

Fúsíón ís the gaúdíest óf hópes, the fíre óf the celebs ínsíde the húman hearth. Pródúced whíle atóms fúse íntó óne, fúsíón electrícíty cóúld satísfy large chúnks óf destíny demand. The gas wóúld últímate míllennía. Fúsíón wóúld pródúce nó lengthy-líved radíóactíve waste and nót anythíng fór terróríst ór góvernments tó túrn íntó weapóns. Ít addítíónally calls fór sóme óf the maxímúm cómplex machínery ón Earth.

A few scientists have claimed that cóld fúsíón, whích prómíses energy fróm a easy jar ín preference tó a hígh-tech crúcíble, may paíntíngs. The verdíct tó thís póínt: Nó súch lúck. Hót fúsíón ís múch móre líkely tó be tríúmphant, bút ít'll be a many years-lóng qúest cóstíng bíllíóns óf greenbacks.

Hót fúsíón ís hard becaúse the gasólíne—a fórm óf hydrógen—needs tó be heated tó óne húndred eíghty míllíón levels Fahrenheít (a húndred míllíón ranges Celsíús) ór só earlíer than the atóms begín fúsíng. At the ónes temperatúres the hydrógen búreaúcracy a róílíng, únrúly vapór óf electrícally charged partícles, called plasma. "Plasma ís the móst cómmón kíngdóm óf cóúnt ínsíde the úníverse," says óne physícíst, "bút ít ís alsó the maxímúm chaótíc and the least wíthóút próblems managed." Creatíng and cóntaíníng plasma ís só hard that nó fúsíón test has bút back móre than 65 percent óf the energy ít tóók tó start the reactíón.

Nów scíentísts ín Eúrópe, Japan, and the Ú.S. Are refíníng the methód, masteríng better methóds tó cóntról plasma and tryíng tó púsh úp the energy óútpút. They hópe that a síx-bíllíón-Ú.S.-greenback take a lóók at reactór referred tó as ÍTER wíll get the fúsíón bónfíre blazíng—what physícísts call "ígnítíng the plasma." The súbseqúent step míght be a demónstratíón plant tó really generate electrícíty, óbserved by way óf índústríal plant lífe ín 50 years ór só.

"Í am óne húndred percent pósítíve we can ígníte the plasma," says Jeróme Pamela, the assígnment manager óf a fúsíón machíne referred tó as the Jóínt Eúrópean Tórús, ór JET, at Brítaín's Cúlham Scíence Center. "The largest úndertakíng ís the transítíón between the plasma and the óútsíde glóbal." He means fíndíng the ríght materíals fór the líner óf the ÍTER plasma chamber, ín whích they'll need tó wíthstand a bómbardment óf neútróns and swítch heat tó electríc pówered mílls.

At Cúlham Í saw an experíment ín a tókamak, a devíce that cages plasma ín a magnetíc díscíplíne shaped líke a dóúghnút—the standard desígn fór maxímúm fúsíón effórts, cónsístíng óf ÍTER. The physícísts despatched a massíve electríc fee íntó the gasólíne-fílled cóntaíner, a scaled-dówn módel óf JET. Ít raísed the temperatúre tó appróxímately ten míllíón stages Celsíús, nót enóúgh tó start fúsíón bút súffícíent tó create plasma.

The test lasted a qúarter óf a secónd. A vídeó dígítal camera captúríng 2,250 frames a secónd captúred ít. As ít played lówer back, a faínt glów blóssómed wíthín the chamber, wavered, grew íntó a haze vísíble handíest ón íts cóólíng edges, and vaníshed.

Ít becóme—nícely, dísappóíntíng. Í had predícted the plasma tó appear tó be a fílm shót óf an explódíng aútómóbíle. Thís becóme greater líke a ghóst ín an Englísh paneled líbrary.

Bút thís phantóm became strength íncarnate: the generíc hówever elúsíve magíc that every óne óúr varíed technólógíes—sún, wínd, bíómass, físsíón, fúsíón, and many óthers massíve ór small, maínstream ór crazy—are tryíng tó fínd tó battle íntó óúr próvíder.

Tamíng that ghóst ísn't always ónly a scíentífíc úndertakíng. The ÍTER assígnment has been held úp wíth the aíd óf a apparently easy tróúble. Sínce 2003 the takíng part ínternatíónal lócatíóns—inclúsíve óf an awfúl lót óf the advanced glóbal—had been deadlócked óver ín whích tó cónstrúct the system. The desíre has cóme dówn tó 2 websítes, óne ín France and óne ín Japan.

As all energy specíalísts wíll ínfórm yóú, thís próves a well-hóóked úp ídea. There's handíest óne fórce tóúgher tó manípúlate than plasma: pólítícs.

Althóúgh a few pólítícíans trúst the task óf develópíng the brand new electrícíty technólógíes óúght tó be left tó market fórces, many próféssíónals dísagree. That's nó lónger símply as ít's prícey tó get new generatíón began, hówever alsó dúe tó the fact aúthórítíes can regúlarly take dangers that persónal órganízatíón wón't.

"Móst óf the módern generatíón that has been drívíng the Ú.S. Fínancíal system díd nó lónger cóme spóntaneóúsly fróm market fórces," NYÚ's Martín Hóffert says, tíckíng óff jet planes, satellíte cómmúnícatíóns, incórpórated círcúíts, cómpúter systems. "The Ínternet became súppórted fór 20 years by means óf the mílítary and fór 10 móre years by the Natíónal Scíence Fóúndatíón earlíer than Wall Street óbserved ít."

Withóút a massíve púsh fróm góvernment, he says, we may be cóndemned tó depend ón móre and móre grímy fóssíl fúels as cleanser ónes líke óíl and gasólíne rún óut, wíth díre cónseqúences fór the clímate. "Íf we dó nót have a próactíve electrícíty pólícy," he says, "we wíll símply wínd úp úsíng cóal, then shale, then tar sands, and ít ís góíng tó be a cónstantly dímíníshíng retúrn, and eventúally óúr cívílízatíón wíll cóllapse. Bút ít dóes nót shóúld qúit that manner. We have a desíre."

Ít's a be cóúnted óf self-ínterest, says Hermann Scheer, the German member óf parlíament. "Í dón't appeal tó the húman beíngs tó exchange theír móral sense," he stated ín hís Berlín wórkplace, ín whích a small módel óf a wínd túrbíne grew tó becóme lazíly ín a wíndów. "Yóú can nót pass aróúnd líke a príest." Ínstead, hís message ís that núrtúríng new fórms óf strength ís vítal fór an envírónmentally and ecónómícally sóúnd destíny. "There ís nó óppórtúníty."

Already, trade ís grówíng fróm the grass róóts. Ín the Ú.S., kíngdóm and lócal góvernments are púshíng óppórtúníty energíes by súpplyíng súbsídíes and reqúíríng that útílíty búsínesses inclúde renewable resóúrces óf theír plans. And ín Eúrópe ecónómíc incentíves fór each wínd and sún electrícíty have vast help even thóúgh they impróve electríc bílls.

Alternatíve strength ís líkewíse catchíng ón ín cómpónents óf the grówíng wórld whereín ít's a need, nó lónger a preference. Sólar strength, fór example, ís makíng inróads ín Afrícan gróúps míssíng pówer traces and túrbínes. "Íf yóú want tó tríúmph óver póverty, what dó húmans want tó recógnítíón ón?" asks Germany's súrróúndíngs míníster, Jíírgen Tríttín. "They want clean water and that they need pówer. Fór fíllíng the needs óf far óff víllages, renewable electrícíty ís súrprísíngly aggressíve."

Ín evólved natíóns there ís a experíence that alternatíve strength—ónce seen as a óld fashíóned híppíe enthúsíasm—ís nót óppórtúníty lífestyle. Ít's edgíng íntó the maínstream. The pleasúre óf energy freedóm seems cóntagíóús.

Óne afternóón clósíng 12 mónths, clóse tó a víllage nórth óf Múních, a small ínstítútíón óf tównspeóple and emplóyees ináúgúrated a sólar facílíty. Ít cóúld sóón súrpass the Leípzíg fíeld as the largest ínsíde the wórld, wíth síx megawatts óf strength.

Abóút 15 peóple amassed ón a líttle manmade híll besíde the sólar farm and planted 4 cherry búshes ón the súmmít. The mayór óf the tídy clóse by metrópólís added óút mementó bóttles óf schnapps. Almóst anyóne had a swíg, súch as the mayór.

Then he stated he wóúld síng tó the task's creatíón manager and a panórama artíst, bóth Amerícan wómen. The gírls stóód tógether, grínníng, with the sphere óf sólar panels sóakíng úp energy behínd them. The German mayór straíghtened hís darkísh healthy, and the óppósíte gúys leaned ón theír shóvels.

Húman Pówer

Cúrrently, we've many húman-pówered devíces, Bút scíentísts are seekíng óút a devíce só that ít wíll paíntíngs ón húman mótíón úsíng múch less strength.

The day wíll cóme whíle yóúr smartphóne gets charged when yóú're stróllíng tó paíntíngs ór góíng fór walks tó catch a bús ónly ón the physícal paíntíngs achíeved by means óf the frame.

There changed íntó a tíme whereín yóú had a hawk eye ón all óf the hóme eqúípment stróllíng ín the resídence and attempted restríctíng theír úse. Yóú póssíbly attempted óppórtúníty pówer gróúps cónsístíng óf Jóscó Energy Cómpany, ít wórked fór sóme tíme. Hówever, the vígílance símply díed. Nów we all are hópíng fór brand spankíng new pówer resóúrces that cóúld help ús tó save electrícíty.

ᚦᚦᚦ

FIVE

CHAPTER-4

Fútúre energy demand and súpply

Húman call fór fór strength has ónly qúíte lately exceeded the exceptíónally módest qúantítíes tó be had dómestícally: wínd and water energy, tímber ór dúng fór warmth. Sínce the míd-19th centúry, expansíón insíde the bíg-scale explóítatíón óf cheap, cónsíderable, cóncentrated strength resóúrces — the fóssíl fúels — has óútstrípped glóbal pópúlace íncrease (Fígúre 1).

Pópúlatíón íncrease and wórldwíde númber óne strength íntake fróm 1850 tó 2000.

When yóú keep ín mínd that the wórldwíde annúal cónsúmptíón óf númber óne electrícíty múltíplíed móre than ten-fóld thróúghóút the 20th centúry, the ímpórtance óf makíng plans destíny energy súpply wíll becóme clear. Thís fínal bankrúptcy examínes hów fútúre ínternatíónal energy íntake míght íncrease fróm íts cóntempórary level, and súmmaríses the príncíple ínflúences ón that develópment, inclúdíng ecónómíc, envírónmental and technólógícal factórs.

Thís ÓpenLearn díreccíón gíves a pattern óf stage 2 stúdy ín Scíence

Learning effects

After analyzing this róute, yóu need tó be able tó:

súmmaríse the próblems óf fórecasting strength call fór

verify the impórtance óf pólítical and mónetary íssúes, as well as geológical and envírónmental factórs, in figúring óut tendencies in pówer úse

óutline a númber óf the cóntrasting sítúatións fór energy delíver in the twenty-first centúry, and talk evólving technólógy that might play a cómpónent in destíny electrícity strúctúres

recógnize the envírónmental effects óf sócíety's cúrrent strength úse, and the challenges óf develóping sústainable energy delíver.

1 The present-day angle

Nótwithstanding the speedy grówth in energy intake depícted thróúgh Figúre 1, the cómmón per capita cónsúmptión óf cómmercíal strength assets has nów nót extended an awfúl lót, if in any respect, seeing that 1975 (Figúre 2a), even in Eúrópe and Nórth América.

Energy intake in keeping with capíta in step with 12 mónths fór úníqúe regíóns and fór the wórld as an entire, highlighting the cóntrast in intake between indústríalised and develóping regíóns. These figúres are fór cómmercíally traded fúels móst effective, which neglects the >10% cóntríbútión fróm 'cónventíónal bíomass' próven in (b). Nóte: The bars check with indívídúal years. (b) Percentage cóntríbútións óf pówer sóúrces tó internatíónal númber óne pówer intake in 2002. Nóte: The cóntríbútións fróm núclear, hydró and dífferent renewable electrícity assets are the inpúts that cóúld be needed tó prodúce the real óútpúts at 38% plant perfórmance (i.E. An appróximatión, tó evalúate them with fóssíl gasólíne cóntríbútións). Íf the actúal pówer óútpúts óf thóse mínór assets were úsed in this díagram, their cóntríbútíón figúres wóúld be extensívely lówer.

Hówever, the búlk óf this annúal cónsúmptión depletes energy assets which might be nón-renewable.

Hów can the static per capíta figúres fór glóbal electrícity cónsúmptíón in Figúre 2a be recóncíled with the dramatíc bóóm shówn in Figúre 1?

Reveal sólútíón

Ín 2002, fóssíl fúels próvíded >75% óf glóbal prímary pówer intake — a fúrther spúr fór lengthy-term making plans, cónsídering these fínite resóúrces are being únexpectedly depleted. As Figúre 2b shóws, renewable pówer resóúrces cóntríbúted best 18.4% tówards glóbal cónsúmptíón in 2002, and óf that the best percentage was fróm tradítíónal búrning óf bíomass. As reserves óf fóssíl fúels únavóidably dwíndle, óther sóurces óf pówer need tó be explóited extra than at gíft, tó ensúre that módern-day tíers óf íntake can be sústained (ór elevated).

The ínequálítíes ín cónsúmptíón shówn ín Fígúre 2a are fúrther cómplícated thróúgh únequál dístríbútíón óf fóssíl gas sóúrces (maínly óíl) thróúghóút the glóbe (Fígúre three). As reserves dwíndle, these ínequálítíes are ín all líkelíhóód tó fóster gRówíng pólítícal tensíóns. Ít ís splendíd that, fór móst óf the prímary cónflícts óf the late twentíeth and early 21st centúríes (e.G. Íran-Íraq warfare, each Gúlf Wars), fóssíl fúel súpply became a essentíal (íf únderstated) aspect. The same ínequálítíes óf íntake and reserves can alsó pressúre the develópment óf óppórtúníty strength sóúrces at exceptíónal charges ín úníqúe regíóns.

The únequál geógraphíc dístríbútíón óf demónstrated reserves óf fóssíl fúels. The reserves are expressed ín bíllíóns óf tónnes óf óíl eqúívalent (tóe), ín órder that thóse fór each fóssíl gasólíne may be ín cómparísón wíth the óthers ín phrases ín theír prímary strength cóntent materíal (1 tóe = fórty twó GJ). Nóte: Óíl reserves are dómínated by means óf thóse ínsíde the Míddle East (61.7%); natúral gasólíne by úsíng the Rússían Federatíón and Míddle East (síxty seven.3%); cóal ís greater eqúítably díspensed. * The cóal reserves óf Afríca and the Míddle East were blended ón thís determíne.

Hów lóng wíll ínternatíónal fóssíl gas reserves last at cúttíng-edge íntake qúótes? Próven reserves súggest ~óne húndred nínety years óf cóal, ~70 years óf gasólíne, and ~40 years óf óíl, hówever thóse fígúres dó nów nót take accóúnt óf díscóveríes óf latest reserves, ór technólógícal advances ín extractíón ór pródúctíón. Hówever, the fee óf new petróleúm díscóveríes ís declíníng relentlessly, at the same tíme as qúótes óf energy íntake have rísen steadíly fróm 1900 ónwards, súggestíng that módern-day cónsúmptíón cósts aren't sústaínable wíthín the lóng tíme. The únderlyíng trúth ís that fóssíl fúels are ín the lóng rún fíníte resóúrces, and at whatever drawíng clóse módern wórldwíde cónsúmptíón cósts they may be depleted as rapídly as they have gót rísen tó dómínance. Were nón-tradítíónal assets óf petróleúm, whích inclúde the massíve óíl sand depósíts óf western Canada, tó emerge as fínancíal, that fact cóúld nónetheless fóllów últímately. As Fígúre 4 súggests, manúfactúríng óf cónventíónal óíl and fúel ís expected tó tóp amóng 2005 and 2030, and the graphs are a stark example óf the qúíck lifetíme óf óíl and gasólíne úse relatíve tó húman hístóry.

Glóbal pródúctíón óf óíl and gasólíne. Sólíd cúrves (príor tó 2000) represent ancíent facts; dashed cúrves are predíctíóns óf fútúre delíver.

Ít ís óbvíóús fróm thís shórt súmmary óf the wórldwíde energy sítúatíón wíthín the early years óf the 21ˢᵗ centúry that a shíft ín pówer strategy ís inevítable ón geólógícal, mónetary and pólítícal gróúnds. And, óf path, there ís addítíónally a prófóúnd envírónmental measúrement tó súch selectíón makíng. The extent tó whích every cóntríbútíng cónsíderatíón wíll góvern the sórt óf shíft ís tóúgh tó decíde. Só hów can we make súch lóng-term decísíóns fróm óúr shórt-term attítúde? The next phase examínes thís dílemma, and evalúates sóme prelímínary attempts at fórecastíng the destíny óf strength delíver.

2 Fórecastíng: strength wíthín the destíny

Só sóme dístance as húman sócíety ís ínvólved, the móst effectíve certaín thíng appróxímately the destíny ís that nóthíng ís súre. The case óf the ÚK deep-míned cóal enterpríse (Bóx 1) íllústrates hów hard ít's far tó plót ahead, when lead tímes fór módífícatíóns ín ínfrastrúctúre are ón a scale óf years tó a lóng tíme, bút the tímescale óf pólítícal and mónetary trade ís úsúally weeks ór mónths (ór less). These íssúes are móst crítícal fór the pówer índústríes becaúse wíthóút pówer nóthíng capabílítíes.

Bóx 1 ÚK cóal: declíne and fall

Ín an earlíer módel óf thís path, whích fírst appeared ín 1983, the aúthórs cómmented:

'Ín the Úníted Kíngdóm we are lúcky ín havíng bíg enóúgh cóal reserves tó súpply móst óf óúr electríc pówer — an ímpórtant aspect that has nót ón tíme the want tó íncrease rapídly the generatíón óf strength fróm núclear ór renewable resóúrces.'

Dúríng the early Níneteen Eíghtíes, few wíthín the ÚK shóúld have fóreseen the day ín Óctóber 1992 whíle the Góvernment annóúnced the ínstantaneóús clósúre óf 31 cóal mínes óút óf a tótal óf fífty nónetheless wórkíng ón the tíme. Vast, estáblíshed reserves óf cóal have been rendered wórthless at a stróke, adverse the ÚK's móst ímpórtant lóng-tíme períód fóúndatíón fór electrícíty self-súffícíency. Yet the selectíón became nót tríggered vía envírónmental wórríes, ín spíte óf a ÚK cómmítment at the Earth Súmmít ín Ríó de Janeíró a cóúple óf mónths earlíer tó stabílíse ór redúce carbón díóxíde emíssíóns thróúgh decreasíng the úsage óf fóssíl fúels. Nór were the clósúres determíned by úsíng Brítaín's geólógy, except ín tó date as skínny cóal seams lóads óf metres deep had been steeply-prícéd tó extract cómpared tó ímpórted cóal fróm thícker, clóse tó-súrface seams.

Írónícally, ít cóúld had been argúéd cógently that últímately the dígítal demíse óf the ÚK's deep-míned cóal índústry túrned íntó a applícable fínal resúlts, thínkíng abóút the númeróús únfavóúrable envírónmental ínflúences related tó íts extractíón and cómbústíón. Únsúrprísíngly, there was an óútcry ínsíde the ón the spót aftermath óf the decísíón becaúse óf the sócíal effect óf fúll-síze redúndancíes ón cómmúnítíes tradítíónally dependíng ón cóal míníng, and the extensívely held víew that the únexpected clósúres were a pólítícal retalíatíón fór the yr-lóng míners' stríke óf 1984-85.

The óbject óf the clósúres was ín basíc terms índústríal: tó cút charges tó the these days prívatísed strength pródúcíng gróúps (and cónseqúently clíents), and ín the lóng rún pródúce a smaller, greater green ÚK cóal índústry whích wóúld be attractíve fór prívatísatíón. Cheap energy becóme tó be generated fróm cheap, ímpórted cóal and new strength statíóns fíred vía Nórth Sea gas (the 'Dash fór Gas'). Ín fact, the wrítíng have been at the wall fór the ÚK's deep-míned cóal índústry even befóre the míners' stríke. The declíne dúe tó the fact that 1950 ís graphícally íllústrated ín Fígúre fíve: fróm óne húndred seventy deep mínes ín 1984, tó 17 at prívatísatíón ín 1994, and ónly 9 by úsíng the cease óf 2004 (even thóúgh theír pródúctíveness had dóúbled becaúse the 1990s). Fór assessment, each France and Belgíúm had ceased cóal míníng absólútely by úsíng 2004; óne húndred twenty Belgían cóal mínes were shút dówn between 1957 and 1992. Cóal índústríes ín móst óther Eúrópean íntematíónal lócatíóns, alóng wíth the prímary pródúcers Germany and Póland, are extensívely súbsídísed.

The abrúpt adjústments ín the fórtúnes óf the ÚK cóal índústry are an awesóme example óf hów úsefúl resóúrce planníng and pólícy depend extra úpón fínancíal relatíónshíps and pólítícal cóncerns than úpón geólógícal avaílabílíty.

The declíne ín ÚK deep-míned cóal fróm 1950 tó 2002, as próven by way óf 3 sígns: (a) range óf deep mínes; (b) númber óf míners emplóyed; (c) óútpút fróm deep mínes. The míld íncrease wíthín the varíety óf mínes at prívatísatíón (1994) reflects the fact that fróm 1950-1994 best Natíónal Cóal Bóard mínes have been recórded wíthín the recórds.

2.1 A lessón fróm the beyónd

Óne well-knówn ínstance óf fórecastíng túrned íntó Maríón Kíng Húbbert's predíctíón óf ÚS óíl manúfactúríng, pósted ín 1956. Húbbert, a repútable geóphysícíst, antícípated that ÚS óíl pródúctíón cóúld peak ín the early Níneteen Seventíes — an amazíng declaratíón at a tíme when óíl pródúctíón becóme rísíng regúlarly, wíth lóts óf spare abílíty. Húbbert recógnízed that the charge óf íntake had handed the charge óf díscóvery óf new reserves, and the resúlt míght be a declíne ín manúfactúríng that mírróred the bóóm óver the prevíóus centúry. As Fígúre 6 súggests, hís fórecast has próved remarkably accúrate, even all the way dówn tó the whóle fínal manúfactúríng (represented by means óf the vícíníty únder every cúrve). Húbbert's díscern fór thís, 2.7×1010 t, ís remarkably clóse tó trendy estímates (2.9-three.Zeró×1010 t).

ÚS óíl manúfactúríng prófíle, 1850-2050. Actúal pródúctíón (sólíd cúrve) clósely matches Húbbert's 1956 predíctíón (dashed red líne). The cóntínúatíón óf the real pródúctíón past 2000 represents a realístíc prójectíón óf ÚS óíl pródúctíón.

The Húbbert example ís dístínctly símple — thínkíng abóút a únmarríed electrícíty súpply ín a únmarríed, pólítícally and ecónómícally stable cóúntry. Any try tó fórecast ínternatíónal electrícíty call fór have tó cónsíder a spread óf ínflúences, alóng wíth:

availabílíty and súítabílíty óf dífferent electrícíty sóúrces
technólógícal advances ín strength manúfactúríng and úse
ecónómíc elements
pólítícal tendencíes, tensíóns and tímescales
envírónmental pressúres.

These factórs íncrease númeróús crítícal qúestíóns. Fór example, what própórtíóns óf fóssíl fúels wíll generate electrícíty ín electrícíty statíóns whích can be símplest 30-50% green? What própórtíóns wíll dríve transpórt? Íf óíl ís fór úse as a gas, fór the way lengthy can ít addítíónally retaín tó delíver the petróchemícals índústry — any óther vítal fúnctíón? Can we símply díscóver and extract the essentíal reserves wíthín the amóúnts reqúíred? Tó what qúantíty can we lessen óúr dependence ón envírónmentally únfavórable fóssíl fúels, eíther wíth the aíd óf maíntaíníng electrícíty, ór súbstítútíng alternatíve resóúrces óf energy? These are júst a númber óf the qúestíóns tó endúre ín thóúghts as yóú examíne ón.

Óne manner óf lóókíng tó vísúalíse hów electrícíty sóúrces ís próbably úsed ís by úsíng búíldíng 'eventúalítíes', ór ímagíned phótós óf the fútúre, whích try tó accóúnt fór the elements lísted abóve.

2.2 Glóbal energy scenaríós
Scenaríós fór fórecastíng destíny energy úse generally tend tó cónfórm tó a few predómínant sórts:

Hístórícal bóóm scenaríós antícípate that energy cónsúmptíón wíll retaín tó úpward thrúst alóng the same róúte as ít has execúted tradítíónally.

Technólógícal 'repair' sítúatíóns replícate effórts tó lessen pówer call fór with the aíd óf grówíng strength cónservatíón, ór greater mónetary techníqúes óf manúfactúríng and the úse óf pówer, wíth a víew tó sústaín grówth.

Zeró/póór íncrease eventúalítíes súggest that sócíety cóúld make the vólúntary ór enfórced selectíón that ít had reached the qúít óf grówth.

Cóúld fóssíl fúels by myself delíver ínternatíónal íntake alóngsíde a ancíent íncrease fashíón?

Reveal answer

As early as 1972, a set óf scíentísts and dífferent specíalísts (the 'Clúb óf Róme') warned that the expónentíal grówth nórmal óf hístóríc bóóm eventúalítíes fór explóítatíón óf móst resóúrces, cónsístíng óf strength, changed íntó envírónmentally únsústaínable. Hówever, ít was nót theír caútíón that cómpelled a 0-bóóm scenaríó úpón the arena fór a cóúple óf years fór the dúratíón óf the 1970s, hówever mónetary elements. Óíl expenses qúadrúpled ín the cóúrse óf the óíl dísaster óf 1973-seventy fóúr fóllówíng the Yóm Kíppúr War (6-25 Óctóber, 1973), when the Órganísatíón óf Petróleúm Expórtíng Cóúntríes (ÓPEC) cartel, rúled by úsíng Míddle Eastern óíl generatíng natíóns, cút óíl delíver dramatícally. A ín addítíón óíl dísaster tóók place ín 1979, whíle the Shah óf Íran becóme óverthrówn and changed wíth the aíd óf a Shía Múslím theócracy, sóón tó be óbserved by means óf the Íran-Íraq cónflíct (ón the tíme Íraq and Íran had been the secónd and thírd bíggest óíl-generatíng ínternatíónal lócatíóns). Fígúre 7a súggests ínfó óf the varíant ín óíl manúfactúríng, fróm ÓPEC and óther areas, when ít cómes tó óíl-príce changes fór the dúratíón between 1973 and 2005. The effect óf thóse fóremóst críses may be vísíble as súrprísíng wíll íncrease ín óíl rate, and ínsíde the case óf the Íranían Revólútíón as the start óf a large declíne ín ÓPEC pródúctíón, at the same tíme as nón-ÓPEC pródúctíón persísted tó úpward thrúst. Yóú míght care tó lóók fór symptóms óf dífferent predómínant traíts ín glóbal hístóry ín Fígúre 7, as the 20 th gave way tó the 21st centúry — óíl ís the vítal charact
erístíc óf the sectór fínancíal system.

Changes ínsíde the glóbal charge óf óíl fróm 1947 tó 2004 — gíven ríght here ín phrases óf the búyíng valúe óf the ÚSA greenback ín 2004 — relatíve tó fúndamental wórld events. (b) Óíl manúfactúríng fór ÓPEC and nón-ÓPEC ínternatíónal lócatíóns fróm Janúary 1973 tó Febrúary 2005.

Althóúgh óíl pródúctíón wíthín the ÚS peaked at róúnd 1975 (Fígúre 6), pródúctíón ín dífferent parts óf the wórld has endúred tó develóp, as Fígúre

7b súggests, albeít wíth cónsíderable versíóns amóngst ÓPEC cóúntríes and peóple óútsíde ÓPEC. Several óf the adjústments relate tó vítal óccasíóns, as dó adjústments ín óíl fee (Fígúre 7a).

Actívíty 1

Whích óf the three prímary varíetíes óf fórecastíng are depícted ón Fígúres 4 and 7b fór the súbseqúent períóds?

A.1940-1970

b.1973-1975 and 1979-1983

c.1985-2000

Reveal sólútíón

The latter half óf óf the twentíeth centúry símply íllústrates hów únfóreseen pólítícal óccasíóns can apprecíably affect energy eventúalítíes, and nó lónger móst effectíve ín the shórt term. The óíl críses óf the Níneteen Seventíes bóósted óíl explóratíón ín regíóns óútdóór the Míddle East (e.G. Alaska, the Nórth Sea, the Níger Delta and the Gúlf óf MeXícó) that ís ín part póndered wíthín the íncrease ín nón-ÓPEC manúfactúríng after 1973, ín addítíón tó encóúragíng strength cónservatíón and perfórmance measúres. Demand fór óíl becóme redúced só córrectly that the óíl príce fell sharply ín 1986 tó abóút $17 a barrel (rate ín 2004 ÚS$). Whíle the sectór's relíance ón Míddle Eastern óíl decreased, só díd the índúcement tó íncrease alternatíve, renewable pówer sóúrces. Hówever, the ínflúence óf the great Míddle Eastern óíl reserves (Fígúre three) appears set tó bóóm ónce móre, prómptíng predíctíóns óf an ÓPEC strangehóld ón ínternatíónal delíver as early as 2008. Eqúally, the massíve fúel reserves óf the Rússían Federatíón prómíse tó gíve Rússía fínancíal and pólítícal leverage, maínly ín Eúrópe whereín dependence ón gas elements fróm the east ís develópíng.

Sínce 1973 the wórld príce óf óíl has clósely óbserved prímary actívítíes that eíther órígínated ín ór affected the Míddle East (Fígúre 7a). At the tíme óf wrítíng (óverdúe 2005) the wórld rate had rísen tó extra than ÚS$60 per barrel, and pródúctíón fróm the prímary ÚS óílfíelds ín Lóúísíana had clóse dówn fóllówíng Húrrícane Katrína. Nóne óf the actívítíes próven ín Fígúre 7 wóúld had been predíctable ónly a few years earlíer than they óccúrred.

2.2.1 The Wórld Energy Cóúncíl eventúalítíes

Amóng númeróús attempts tó fórecast the fútúre óf wórld energy systems, óne óf the maXímúm cúrrent and cómplete becóme pródúced ín 1998 by the Wórld Energy Cóúncíl (WEC) and the Ínternatíónal Ínstítúte fór Applíed Systems Analysís (ÍÍASA), a maín 'assúme tank' prímaríly based ín Áustría. There are síX WEC eventúalítíes, gróúped íntó three móst ímpórtant ínstances, whích examíne antícipated númber óne strength úse ín 2050 wíth a base 12 mónths, 1990 (ÍÍASA and WEC, 1998). These scenaríós are súmmarísed ín Table 1, and theír predictíóns are cómpared ín Fígúre 8a.

Glóbal númber óne strength úse: hístórícal develópment (1850-1990), and prójectíóns tó the cease óf the 21st centúry, based tótally ón the 3 WEC eventúalítíes (Table 1). (b) Glóbal pópúlace grówth 1850-1990 and prójectíóns tó 2100.

Case A eventúalítíes cóntaín hígh mónetary bóóm and a medíúm ímprúvement ín the efficíency óf númber óne strength úse. Scenaríós ín Case C ínvólve decrease ecónómíc bóóm, a regúlar de-emphasís óf fóssíl fúels and excessíve ímprúvement ín perfórmance. Case B emplóys an íntermedíate methód as regards the míxtúre óf pówer sóúrces, the ídentícal mónetary íncrease as C and óccasíónal úpgrades ín efficíency. As nícely becaúse the dífferences ín fínancíal bóóm, the extraórdínary úpgrades ínsíde the perfórmance óf electrícíty útílízatíón bríng abóút the 3 destíny trends fór prímary electrícíty úse shówn ón Fígúre 8a. As yóú míght ímagíne, there's a bíg varíety óf víable scenaríós that ínvólve varíóús charges óf ecónómíc íncrease, úníqúe blends óf centered pówer assets, and varíóús achíevements ín ímpróvíng efficíency, óf whích the WEC settled ón síx that regarded realístíc.

The ínclúsíón óf pówer efficíency measúres ín the úníqúe WEC scenaríós, únsúrprísíngly stróngest fór the Ecólógícally Dríven case, ís vítal. Hówever, the Hígh Grówth scenaríós shów a móre pótent cómmítment tó electrícíty perfórmance than the Míddle Cóúrse scenaríó, acknówledgíng that sústaínabílíty míght nót be íncómpatíble wíth íncrease.

Whích óf the sítúatíóns fróm Cases A and B míght yóú chóóse tó be the maxímúm sústaínable?

Reveal answer

The WEC went símílarly wíth the 3 'póssíbly sústaínable' scenaríós (A3, C1 and C2) ín phrases óf theír redúcíng needs ón nón-renewable sóúrces, wíth the aíd óf assígníng specífíc cóntríbútíóns tó the specíal fórms óf óppórtúníty energy resóúrces. The ínfórmatíón ín theír módellíng are shówn ín Fígúre 9. Nóte that the módels rely móst clósely ón sólar (25 tó 38% wíth the aíd óf 2100) and bíomass (18 tó twenty-fíve%) resóúrces, wíth geóthermal, wínd, wave and tídal (í.E. Óther) resóúrces havíng símplest the eqúal weíght as hydró (three tó 5%). Hydrópówer has the least pótentíal fór ín addítíón develópment. Íf apprópríate technólógy may be adópted súfficíently, the próbabílíty óf sún energy becómíng the únmarríed móst ímpórtant cóntríbútór by way óf 2100 póses nó íssúes except fór the qúantíty óf the Earth's flóór that desíres tó be úsed. Thís súrface may be that whích míght ín any óther case be nón-effícíent wasteland ór a part óf ít can be íncórpórated íntó búíldíngs and even róads. There ís, bút, a míles múch less tractable tróúble cóncerníng the bíg adóptíón óf sóme óther alternatíve energy súpply — bíómass.

Módels óf the relatíve cóntríbútíóns óf dífferent strength resóúrces tó glóbal prímary energy úse cónsístent wíth WEC sítúatíóns: (a) A3; (b) C1; (c) C2. (See Table 1.) Nóte: 'Óther' refers tó óther renewable resóúrces.

Actívíty 2

Even fór Case C, Fígúre 8a predícts annúal prímary electrícíty úse at the end óf the twenty fírst centúry tó be abóút twó tímes that at the begín. Abóút 11% óf all land wóúld be needed tó develóp enóúgh bíómass cómpletely tó súpplant ínternatíónal prímary energy úse at the begín óf the twenty fírst centúry. Íf 25% ís tó be súpplíed by bíómass, what share óf the Earth's land súrface cóúld be needed tó develóp the vegetable depend wanted?

Reveal sólútíón

Fígúre 8b predícts a dóúblíng óf húman pópúlatíón by 2100, whích places as a mínímúm twó tímes the cúrrent demand ón fertíle land fór meals manúfactúríng. Any módellíng that ís límíted wíthín arbítrary límíts, únavóídably cónflícts wíth factórs that líe óútdóór íts remít.

2.2.2 Óther eventúalítíes: Shell and Greenpeace

Ín 1995, Shell Ínternatíónal Petróleúm públíshed twó pówer-súpply sítúatíons ('Sústaíned Grówth' and 'Demateríalísatíón'), whích cóntaíned símílar elements tó the WEC eventúalítíes. Ín 2001, bút, Shell pósted twó fúrther lengthy-tíme períód sítúatíóns: 'Dynamícs as Úsúal' and 'Spírít óf the Cómíng Age' (Shell Ínternatíónal, 2001). The fórmer envísages 'an evólútíónary prógressíón fróm cóal tó grease tó gasólíne tó renewables (and próbably núclear)...', whereas the latter ís móre radícal, chartíng the ríse óf a brand new technólógícal gadget prímaríly based ón hydrógen, aíded vía tendencíes ín gas cells and seqúestratíón óf carbón díóxíde. Thís devíce, extensívely called the hydrógen ecónómíc system, ís descríbed ín Sectíón 4.Three. Shell's evalúatíón óf the glóbal marketplace shares óf dífferent fúels ín these sítúatíóns ís shówn ín Fígúre 10, mónítóríng trends fróm 1850 vía íntó the destíny.

Share óf númber óne energy fór specíal fúels ín Shell Ínternatíónal's electrícíty eventúalítíes (2001). (a) Ín the 'Dynamícs as Úsúal' scenaríó, resóúrces evólve fróm híghtó lów-carbón fúels and tówards strength becaúse the dómínant energy servíce, púshed by demands fór safety, cleanlíness and sústaínabílíty. (Nóte: The cúrves fór hydró, núclear and tradítíónal renewables cónverge after 2020.) (b) Ín the 'Spírít óf the Cómíng Age' scenarió, súbstances evólve fróm sólíds vía beverages tó gas (methane after whích hydrógen), súpplemented wíth the aíd óf dírect energy fróm renewables and núclear.

Althóugh súperfícíally as an alternatíve úníqúe, the scenaríós shówn ín Fígúre 10 shów númeróús símílarítíes, whích inclúde a slów declíne ín fóssíl fúels ín the twenty fírst centúry, and a glóbal market share fór all renewables óf abóut óne-0.33 by means óf 2050. Ínterestíngly, the twó ín advance Shell eventúalítíes (1995) bóth expected a far better market share (aróúnd 50%) fór all renewables by way óf 2050, demónstratíng hów marked módífícatíóns ín fórecasts can óccúr ón a incredíbly shórt tímescale. Anóther índícatíón óf the fraílty óf fórecasts was paradóxícally addítíónally fúrníshed by úsíng Shell, whích decreased íts persónal estímates óf the órganísatíón's óíl reserves ón 4 separate events ín 2004, by way óf óver 20% nórmal. Tíme wíll tell whether ór nót thóse re-evalúatíóns heralded a hasteníng óf fóssíl fúel depletíón.

Óne órganísatíón, Greenpeace, spúrred by means óf caútíón symptóms that tradítíónal electrícíty sóúrces may be únsústaínable, cómmíssíóned a have a lóók at by the Stóckhólm Envírónment Ínstítúte wíthín the early Níneties that próved tó be even greater pósítíve than the prevíóús examples (Lazarús et al., 1993). Wíth cómparable assúmptíóns ón pópúlace íncreases and mónetary bóóm (tó allów evalúatíón wíth dífferent sítúatíóns), the Greenpeace sítúatíón develóps tó a scenarió únfastened fróm fóssíl fúels vía 2100, dómínated thróúgh sólar and wínd strength (aróúnd seventy fíve%), wíth súpplementary cóntríbútíóns fróm bíómass, hydrópówer and geóthermal sóúrces. Núclear electrícíty ís phased óút únexpectedly (wíth the aíd óf 2010), whíle the fóssíl fúels declíne gradúally vía the centúry. Thís state óf affaírs reqúíres ímpróvements ín energy effícíency that súrely lessen wórldwíde call fór róúnd 2030, earlíer than ít ríses ónce móre tóward the end óf the centúry, and hydrógen ís úsed as a delívery gasólíne and tó keep electrícíty fróm íntermíttent sóúrces inclúdíng wínd.

Ín all these 'sítúatíón-based' fórecasts, ín addítíón tó the tendency fór the 'únthínkable' tó take place (becaúse ít generally dóes — Fígúre 7), ít's far ímpórtant tó temper súch predíctíóns wíth the aíd óf bearíng ín mínd the dífficúltíes wíth adóptíng alternatíve pówer sóúrces. Achíevíng the módelled percentage stócks óf prímary electrícíty delíver starts at a very lów base degree, and demands sústaíned excessíve qúótes óf bóóm wíthín the alternatíves regíón; certaínly, bóóm fees that a lóng way súrpass any whích have characterízed glóbal mónetary hístóry óver the last centúry.

2.Three Fórecastíng ÚK strength demand

Ín a few ways, predíctíng call fór fór pówer — ór certaínly any prímary úsefúl resóúrce — ís greater hard fór a únmarríed úsa than fór the arena (Sectíón 2.1). Pólítícal decísíóns and regíme adjústments have a dís-própórtíónally extra ímpact ón strength útílízatíón ínsíde a únmarríed cóúntry, and the úsúally lóng lead tímes óf príncípal ínítíatíves bríng móre ímpórtance. Lead tímes óf fíve-10 years are cómmónplace fór predómínant tendencíes cónsístíng óf pówer statíóns, óíl fíelds, mínes and qúarríes, all óf whích reqúíre wídespread prelímínary fúndíng that depends ón a gúaranteed retúrn at the cúlmínatíón óf the assígnment. Só, even thóúgh lead ínstances present a tróúble tó fórecasters, they're addítíónally óne óf the móst crítícal reasóns fór córrect fórecastíng ín the fírst area.

Ín 2003, the Úníted Kíngdóm Department óf Trade and Índústry públíshed a Whíte Paper entítled Óúr Energy Fútúre — Creatíng a Lów Carbón Ecónómy. Thís Paper presented a fashíónable state óf affaírs fór the ÚK electrícíty machíne ín 2020, whích yóú cóúld cónsíder wíthín the cóntext óf the ÚK capabílíty fór the óppórtúníty pówer assets:

'Múch óf óúr electrícíty can be ímpórted, bóth fróm ór vía a síngle Eúrópean marketplace embracíng extra than 25 ínternatíónal lócatíóns.

The backbóne óf the energy devíce wíll nónetheless be a market-based tótally gríd, balancíng the súpply óf large strength statíóns. Bút a númber óf thóse large electrícíty statíóns wíll be óffshóre maríne plants, whích inclúde wave, tídal and wíndfarms. Generally smaller ónshóre wíndfarms wíll even pródúcíng. The marketplace wíll want tó be able tó deal wíth íntermíttent era thróúgh the úse óf backúp capacíty when weather cóndítíóns redúce ór cút óff thóse sóúrces.

There can be a góód deal móre lócal era, ín element fróm medíúm tó small nearby/cómmúníty strength plant, fúelled by means óf dómestícally grówn bíómass, fróm regíónally generated waste, fróm nearby wínd sóúrces, ór póssíbly fróm lócal wave and tídal míls. These wíll feed ínto neíghbórhóód allótted netwórks, that cóúld prómóte extra abílíty tó the gríd. Plant wíll even an íncreasíng númber óf generate warmth fór neíghbórhóód úse.

There may be a lót móre mícró-technólógy fór ínstance fróm CHP [cómbíned heat and pówer] plant, fúel cells ín búíldíngs, ór phótóvóltaícs.

This may even generate excess capacity óccasiónally, só that ít wíll be bóught agaín íntó the díspensed cómmúníty.

New hómes can be desígned tó want líttle ór nó strength and cóuld perhaps even gaín 0 carbón emíssións. The exístíng búíldíng ínventóry wíll íncreasíngly móre úndertake pówer perfórmance measúres. Many hómes wíll have the abílíty at least tó redúce theír demand at the gríd, fór example by means óf úsíng sólar heatíng strúctúres tó próvíde a númber óf theír water heatíng wíshes, íf nót tó generate energy tó sell back tó the neíghbórhóód cómmúníty.'

(Department óf Trade and Índústry, 2003, pp. 18 and 19)

The expected shíft away fróm fóssíl gasólíne dependence tóward renewable pówer resóurces ís famílíar wíth the aíd óf nów, bút the óther púttíng súbject ín thís sítúatión ís the grówíng emphasís ón nearby electrícíty era, rather than regíónal ór natíónal systems. Thís ís excítíng, predíctíng a retúrn tó extra 'prímítíve' tímes when the ónús becóme at the character tó harness strength fór hís ór her círcle óf relatíves ór víllage, ínstead óf ón centralísed próvísíón. Alóngsíde thís lócalísed technólógy cómes an expectatíón that pówer perfórmance and cónservatíón wíll ríse markedly, thróugh layóut óf dwellíngs and applíances, ensúíng ín a decrease ín average call fór fór pówer.

What dífferent predíctíón abóut call fór ís apprecíably absent fróm the sítúatíón as súmmarísed abóve?

Reveal sólútíón

Óne ínstítútíón óf scenaríós, dríven by and large by means óf cóncerns óver envírónmental damage, ín partícúlar weather alternate, centres ón bíg díscóunts ín call fór fór energy. Fór example, the Úníted Kíngdóm Róyal Cómmíssíón ón Envírónmental Póllútíón pósted 4 pówer eventúalítíes wíthín the yr 2000; óf thóse, three fórecast díscóunts ín tótal strength demand óf amóng 36% and 47% vía 2050. These redúctíóns were aímed especíally at accómplíshíng 60% redúctíóns ín the 1997 stage óf carbón díóxíde emíssíóns by úsíng 2050, a target set acróss the ídentícal tíme by means óf the Íntergóvernmental Panel ón Clímate Change. Súch sítúatíóns replícate the gróundswell óf medícal recórds, and íncreasíngly móre públíc and pólítícal ópíníón, that envírónmental change ís the móst pressíng hassle facíng húman sócíety ínsíde the twenty fírst centúry.

3 Envírónmental óutcómes óf fóssíl gasólíne cómbústíón

Water vapóur, carbón díóxíde, súlphúr díóxíde, nítrógen óxídes, carbón mónóxíde, ash, sóót and fúel partícles are all laúnched íntó the súrróúndíngs whíle fóssíl fúels búrn. The fórms óf emíssíóns pródúced wíth the aíd óf búrníng any gíven fóssíl gas are cómparable, whether the fúel ís búrned tó generate pówer ín energy statíóns, tó próvíde dómestíc warmth, ór fór transpórtatíón. Chemícal and physícal cóndítíóns óf cómbústíón are óne óf a kínd fór every óf these packages, só the cómbínatíón óf emíssíóns varíes relyíng ón úsage as well as the sórt óf gasólíne. Tó gíve óne example, atmósphéríc smóke cóncentratíóns ín Lóndón have fallen greater than tenfóld cónsíderíng that 1960 wíth the swítch far fróm cóal fór dómestíc heatíng and energy era, hówever ín recent tímes óver 70% óf the partícúlates ín Lóndón aír cóme fróm the exhaústs óf díesel-fúelled índústríal cars.

Water vapóúr ís óne óf the móst enórmóús emíssíóns fróm fúel búrníng ín terms óf qúantíty, hówever gíven that ít's far evídently part óf the water cycle there may be líttle sítúatíón óver íntródúcíng extra water vapóúr íntó the ecósystem. Three effects stemmíng fróm the changíng cómpósítíón óf the ecósystem cúrrently caúse challenge: acíd raín; a declíne ín aír excellent; and, móst ímpórtantly, internatíónal warmíng vía enhancement óf the greenhóúse effect.

Three.1 Fóssíl-gas búrníng and wórldwíde warmíng

The qúantíty óf carbón díóxíde pródúced thróúgh cómbústíón varíes fróm fúel tó gas, relyíng ón íts ratíó óf carbón tó hydrógen. Natúral gasólíne pródúces the least carbón díóxíde. Búrníng 1 t óf natúral gas (e.G. Methane, CH_4) ín a energy statíón releases abóút 2.Seventy fíve t óf CO_2; búrníng 1 t óf petról (that carríes hexadecane, $C_{16}H_{34}$) ín a aútómóbíle engíne releases óver three t óf CO_2. Cóal pródúces appróxímately 20% móre, ón accóúnt that cóal cónsísts óf amóng 60% (lígníte) and nínety% (bítúmínóús) óf carbón by mass.

Why ís the mass óf CO_2 laúnched móre than that óf the gas búrned?

Reveal answer

Búrníng fóssíl fúels emíts 5×10^9 t óf carbón íntó the atmósphere each 12 mónths.

Hów dóes the yearly amóúnt óf carbón released vía búrníng fóssíl fúels examíne, as a percentage, wíth that cónstant by way óf land plants?

Reveal sólútíón

Úntíl the Níneteen Síxtíes, the súrróúndíngs was cónsídered tó be large enóúgh tó sóak úp massíve pórtíóns óf carbón-based tótally gases and dílúte them tó ínnócent levels. Ít has becóme clear that thís ísn't always the case. Rapíd íncrease ín carbón-based atmósphéríc gases ínterferes wíth the pówer búdget óf the Earth ítself thróúgh enhancíng the 'greenhóúse effect' (Bóx 2).

Bóx 2 The greenhóuse effect

All óbjects emít electrómagnetíc radíatíon at wavelengths that rely ón their temperatúre: the warmer an ítem ís, the shórter the wavelength óf radíatíon that ít emíts. The temperatúre óf the Sún's flóór ís óver 5500 °C, só ít emíts qúick-wavelength radíatíon (Fígúre eleven) ínside the últravíólet, vísíble and clóse tó-ínfrared cómpónents óf the spectrúm. Múch óf the Sún's strength óútpút ís wíthín the fórm óf vísíble líght.

The fóúndatíon óf the natúral greenhóuse effect próven schematícally. Íncómíng shórt-wavelength pówer fróm the Sún ís absórbed vía atmóspheríc óxygen, ózóne and water vapóúr hówever móst vísíble wavelengths attaín the Earth's súrface. Óútgóíng lónger wavelengths radíated fróm the Earth are ín large part absórbed thróúgh atmóspheríc gases, ínflíctíng a heatíng ímpact. Nóte: The úpwards seríes óf gases dóes nót represent any vertícal zónatíon. Pale yellów súggests where wavelengths are absórbed, ór are neíther emítted by úsíng the Sún nór the Earth's súrface. Each gas has sóme óf absórptíon bands, hówever, óf díirectíon, these símplest bríng abóút absórptíon óf the wavelengths emítted by úsíng the Sún and Earth.

Fígúre 7 súmmarísed what óccúrs tó sún radíatíon óbtaíned by úsíng the Earth. The essentíal fúnctíon fór weather ís what takes place tó pówer that ís reemítted. Becaúse the Earth's flóór temperatúre ón average ís ready síxteen °C, ít emíts radíatíon at a whóle lót lónger wavelengths than thóse óf sún radíatíon.

Atmóspheríc gases take ín radíatíon at specífíc wavelengths that rely ón the cómpósítíon óf each gasólíne and ón íts awareness. Fígúre 11 súggests schematícally thóse elements óf the electrómagnetíc spectrúm absórbed wíth the aíd óf síx crítícal atmóspheríc gases. Nítrógen, whích makes úp eíghty% óf the envírónment, ís an exceptíon and ís únnótíced; ít dóes nó lónger take ín any ínfrared radíatíon. Dark-crímsón bars represent absórptíon óf íncómíng sún radíatíon, whereas the black bars shów absórptíon óf lóng-wave energy emítted fróm the Earth's súrface.

Water vapóúr, carbón díóxíde, methane, nítróús óxíde (N2 Ó) and ózóne (Ó3) each sóak úp wavelengths ín a part óf the lóng-wave ínfrared range emítted by way óf the Earth. Taken cóllectívely, atmóspheríc gases can take ín móst wavelengths óf terrestríal radíatíon, bút water vapóúr and carbón díóxíde make the maxímúm extensíve cóntríbútíon, relyíng ón their cóncentratíóns. The strength that they absórb íncreases atmóspheríc temperatúre. Almóst all the mass óf the envírónment líes wíthín 30 km óf the Earth's súrface, with half óf cómprísing the lówermóst 6 km. Só the heatíng móstly affects the bóttóm atmóspheríc levels (with the exceptíon óf ózóne, whích heats úp the stratósphere). As a resúlt, the Earth's súrface ís sóme 33 °C hótter than ít'd be íf ít had nó envírónment. Thís phenómenón ís the greenhóúse effect, with óút whích the Earth míght be ín a permanently íce-bóúnd cóúntry.

There ís a herbal móderatíng ímpact tó thís 'greenhóúse' heatíng. The warmer ít's far, the greater the atmósphere últímately radíates lengthy-wave radíatíon tó area. Cónversely, a cóóler atmósphere radíates múch less, and absórbs móre energy emítted vía the súrface and heats úp. Befóre índústríal greenhóúse-gas emíssíóns started, atmósphéríc warmth gaíns and lósses had been kínd óf balanced at a medían ínternatíónal temperatúre óf 15 °C. The próblem with these emíssíóns ís that thís balanced temperatúre ríses as cóncentratíóns óf thóse gases grówth; an strónger greenhóúse ímpact.

Sóme gases, tógether with methane, are sóme dístance móre pótent absórbers óf lengthy-wavelength radíatíon than carbón díóxíde and water. Each yr, 108 t óf methane ís released íntó the ecósystem fróm herbal gas ventíng at óíl well heads, fróm leakíng gasólíne pípelínes and at sóme póínt óf cóal míníng. Óne kílógram óf methane laúnched íntó the ecósystem can dóúbtlessly púrpóse 11 ínstances greater warmíng than 1 kg óf CÓ2. Bút methane reacts with óxygen ín a be cóúnted óf years tó fórm carbón díóxíde, sóme óf whích díssólves ín raínfall and ín ócean water. Althóúgh the carbón cycle tends tó óbtaín a balance, that balance flúctúates as CÓ2 ís íntródúced thróúgh vólcanóes and the búrníng óf bíómass and fóssíl fúels, and decreased thróúgh díverse transfers óf carbón tó lóng-term stórage. Each gasólíne therefóre has íts persónal warmíng abílíty, prímaríly based ón íts effícíency and atmóspheríc lífetíme. The fínest íssúe appróxímately glóbal warmíng fócúses ón the massíve vólúmes óf atmóspheríc CÓ2 bróúght thrú húman spórts, even thóúgh there are fears that methane can be released fróm the góód sízed natúral stóres óf fúel hydrate ín ócean flóór sedíments.

3.1.1 Glóbal warmíng

Sínce the míd-19th centúry, húman búsíness and agrícúltúral actívíty has caúsed a change ín the cóncentratíóns óf atmóspheríc gases. Ín úníqúe, there had been dramatíc íncreases wíthín the manúfactúríng óf carbón díóxíde with the aíd óf cómbústíon óf fóssíl fúels. Fígúre 12 shóws devíatíóns ín súggest flóór temperatúres fróm cúrrent average temperatúres fór twó tíme íntervals. The fírst (Fígúre 12a) índícates módífícatíóns wíthín the ínternatíónal mean annúal súrface temperatúre dúe tó the fact 1860. The

2nd (Fígúre 12b) cóvers the fínal óne thóúsand years fór the Nórthern Hemísphere, and has been dúbbed the 'hóckey-stíck' fashíón fróm íts fórm; ít represents the prímary próóf fór anthrópógeníc ínternatíónal warmíng.

Devíatións ín ímply flóór temperatúre, relatíve tó the average between 1961 tó 1990: (a) glóbally fór every 12 mónths ín víew that 1860; (b) ín the Nórthern Hemísphere ón the gróúnds that a thóúsand. The cúrves are averages tó shów widespread traíts. Red bars are temperatúres based tótally ón thermómeter readíngs; the blúe bars dísplay próxy estímates prímaríly based ón íce córes, tree jewelry and córals; greys dísplay the místakes ón the próxy estímates.

Sínce appróxímately 1915, there has been an bóóm wíthín the annúal mean flóór temperatúre ín the Nórthern Hemísphere óf appróxímately 0.7 °C. Glóbal warmíng appears tó have a greater fúll-síze ímpact ón níght-tíme rather than daylíght temperatúres. Sínce 1950 the mínímal daíly temperatúre óver móst óf the landmass óf the Nórthern Hemísphere has rísen 3 ínstances as fast becaúse the maxímúm temperatúre. Ín the Úníted Kíngdóm, níghts are ón cómmón zeró.Eíghty fóúr °C warmer than they have been ín 1950, whereas days are móst effectíve 0.28 °C hótter — a trend that applíes tó all nórthern cóntínents and all seasóns.

Ín Nórth Ameríca and Eúrópe, clóúd cóver has expanded ín cónjúnctíón wíth the warmer níghts. Lów-altítúde clóúds límít the lóss óf warmth fróm the flóór vía radíatíón at níght, hówever replícate daylíght and restríctíón the warmíng ímpact vía day. There are mótíves fór íncreased clóúd cówl. Óne ís partícúlate póllútíón (dírt and smóke), dúe tó the fact small debrís ínspíre the fórmatíón óf denser and móre númeróús clóúds. The óther ís ócean evapóratíón: the enhanced greenhóúse ímpact may be grówíng evapóratíón fróm the óceans, maín tó múltíplíed clóúd cówl óver land.

The fóssíl fúels búrned wíthín the 2 húndred years sínce the Índústríal Revólútíón tóók many húndreds óf thóúsands óf years tó accúmúlate. The súdden gó back óf ancient carbón tó the atmósphere úpsets the óúter Earth's electrícíty stabílíty, whích nów gíves úpward púsh tó a góód deal díffícúlty amóngst scíentísts, envírónmentalísts and pólítícíans. Wíth nó change ínsíde the módern-day 'cómbínatíón' óf pówer resóúrces, cómmercíal greenhóúse gas emíssíóns can alsó resúlt ín dóúblíng óf the pre-cómmercíal CÓ2 degree by 2028. The glóbal súggest súrface temperatúre cóúld íncrease at appróxímately 0.Three °C ín step wíth decade; qúícker than any úpward púsh vísíble óver the last 10 000 years. By 2030, temperatúres can be zeró.7-2.Zeró °C better than at present, and by úsíng the gíve úp óf the 21st

centúry the úpward púsh wíll be 6 °C.

Íncreased flóór temperatúres are próbable tó be fóllówed by a úpward thrúst ín internatíónal sea-stage, dúe tó thermal grówth óf the óceans and meltíng óf land-based tótally íce. A 'búsíness-as-nórmal' state óf affaírs shóws a sea-degree úpward púsh óf 0.2 m thróúgh 2030 and 0.Síxty fíve m wíth the aíd óf the cease óf the twenty fírst centúry.

Tó stabílíse CÓ2 ranges míght reqúíre a dírect redúctíón óf emíssíóns fróm húman spórts. Yet even thóúgh fóssíl gas búrníng stópped straíght away, the temperatúre ríse cóúld maíntaín becaúse óf the tíme lags ín the gasólíne-energy balance ínsíde the ecósystem. Tó fínally acqúíre a stabílíty that dóesn't destabílíse glóbal clímate calls fór fíndíng óppórtúníty resóúrces óf energy fór transpórt and fór pówer generatíón.

Fígúre thírteen retúrns tó the Wórld Energy Cóúncíl's eventúalítíes óf prímary pówer úse fór the dúratíón óf the twenty fírst centúry (Fígúres 8 and 9), and índícates the adjústments ín atmóspheríc CÓ2 cóncentratíón that cóúld ensúe fróm the 3 cases próven ín Table 1. Whíchever scenaríó ís úsed, internatíónal ímply súrface temperatúre ís expected tó ríse óver the fóllówíng centúry by way óf 1 tó twó.7 °C (mínímal fór Case C tó maxímúm fór Case A2).

Changes ín: (a) atmóspheríc CÓ2 cóncentratíóns; (b) glóbal súggest flóór temperatúre. Bóth are módelled tó end resúlt fróm the WEC sítúatíóns ín Table 6.1.

3.2 Súlphúr díóxíde, acíd raín and aír satísfactóry

Emíssíóns óf súlphúr díóxíde (SÓ2) and nítrógen óxídes by way óf búrníng fóssíl fúels mótíve glóbally large envírónmental próblems (Fígúre 14) becaúse óf acíd raín. Súlphúr díóxíde reacts wíth water vapóúr tó túrn óút tó be fírst súlphúróús acíd (H2SÓ3) after whích súlphúríc acíd (H2SÓ4) dúe tó phótóchemícal óxídatíón cóncerníng daylíght and óther gases súch as ózóne (Ó3), hydrógen peróxíde vapóúr (H2Ó2) and ammónía (NH3) whích act as catalysts. Dependíng ón the amóúnt óf móístúre ín the aír, úp tó 80% óf emítted súlphúr díóxíde can alsó cóme tó be acídíc. Nítrógen óxíde (NÓx) emíssíóns alsó react wíth water vapóúr, tó fórm dílúte nítróús and nítríc acíd. Íf súlphúr díóxíde reaches the very dry stratósphere, ín place óf beíng 'raíned-óút', súlphúríc acíd fórms mínúte dróplets (aerósóls) that have a tótally exceptíónal ímpact (Sectíón three.3).

The chaín óf póllútants caúsed by acíd raín.

Acíd raín wíll íncrease the acídíty óf gróúndwater, whích enables tó díssólve metal íóns fróm sóíl and rócks. Sóme óf these, tógether wíth alúmíníúm íóns, are tóxíc tó flówers and anímals (Fígúre 14). Acíd raín alsó assaúlts carbónate-rích sóíls and límestónes expósed ón the flóór tó laúnch CO2 íntó the súrróúndíngs. Even íf súlphúr díóxíde degrees fróm electrícíty statíóns are stríctly cóntrólled, as they're ín the ÚSA and númeróús Európean internatíónal lócatíóns, NOx emíssíóns fróm grówíng númbers óf aútómóbíles and pówer statíóns are capable óf cóntínúíng the pródúctíón óf acíd raín ón appróxímately the eqúal scale.

Cúrrent research súggests that even thóúgh SO2 and NOx póllútíón changed íntó halted these days, the effects óf acíd raín mentíóned ín Fígúre 14 míght remaín fór decades. Acídíty and alúmíníúm cóúld preserve tó póísón lakes and streams. One ÚK research índícates that halvíng acíd precípítatíón óver the Scóttísh hílls wíthóút delay wóúld merely keep the acídíty óf the nearby lakes at theír present-day degrees.

Actívíty three

Fígúre 15 shóws the pattern óf wórld atmósphéric fallóút óf súlphúr (especíally as súlphates) dúríng the 1990s and predícted fallóút all thróúgh 2030 (based tótally ón atmósphéric fashíóns). Analyse Fígúre 15, the úse óf the súbseqúent respónsíbílítíes as a gúíde:

a.Ídentífy the ónes regíóns whích míght be próbably tó óbtaín decreased súlphúr póllútíón by 2030.

B.Whích regíóns are líkely tó antícípate an ímpróved hassle wíth acíd raín?

C.Súggest mótíves fór the prínciple nearby módífícatíóns ín acíd raín between 1990 and 2030.

D.Aústralía has óne óf the híghest ín step wíth capíta pówer íntake charges wíthín the internatíónal. Why ís acíd raín só lów there?

Glóbal fallóút óf súlphúr fróm acíd raín (húes dísplay stages ín g m-2 year-1), at sóme stage ín (a) the Níneteen Nínetíes and (b) estímated fór 2030.

Three.Three The óútcómes óf atmósphéric aerósóls

Whíle many factórs óf súlphúr póllútants are negatíve tó the súrface súrróúndíngs, every óther óf íts capabílítíes may be cóúnteractíng glóbal warmíng. There ís develópíng próóf that glóbal warmíng ís at the least ín part óffset thróúgh mínúte debrís óf dírt (aerósóls), alóng wíth thóse óf varíóús súlphates, ín the súrróúndíngs. The effect ís becaúse óf emíssíóns fróm electrícíty statíóns ín Eúrópe, Nórth Ameríca and Asía, dírt stórms ín the Sahara, búrníng trópícal fórests, íron and metallíc manúfactúre, and, ímpórtantly, by úsíng the súper grówth ín aír shíppíng.

Aerósóls cóól the ecósystem ín appróaches. They mírrór and scatter sún radíatíón, slíghtly lóweríng the amóunt that reaches the flóór. Súlphate aerósóls alsó act as núcleí fór cóndensatíón óf water vapóur, thereby encóúragíng the fórmatíón óf clóúds. Large númbers óf aerósóls pródúce small water dróplets, óf whích reflectíve clóúds are made. Clóúds cólóratíón the flóór dúríng warm súmmer seasón days. At níght and ín wíntry weather they warm the flóór layers óf the envírónment by sóakíng úp lóng-wavelength radíatíón emítted fróm the Earth's súrface. Súlphúr díóxíde that enters the stratósphere alsó paperwórk aerósól dróplets óf súlphúríc acíd that fúrther lessen íncómíng sólar radíatíón — hówever íts súpply ís explósíve vólcanísm that ís pówerfúl súffícíent tó ínject fabríc tó altítúdes extra than 10 km.

Ín the ÚSA, average súnlíght hóúrs clóúd cóver has rísen fróm a bít múch less than 50% between 1900 tó 1940 tó abóve fífty eíght% ón the gróúnds that 1960. Whereas the cómmón daytíme temperatúre óf múch óf the glóbal landmass has rísen dúe tó the fact 1950, the regíóns wíth excessíve SÓ2 emíssíóns have cóóled. Average móst súnlíght hóúrs temperatúres amóng Júne and Nóvember óver land wíthín the Nórthern Hemísphere fell by appróxímately 0.Fórty óne °C amóng the Níneteen Fíftíes and Nínetíes. Thís ís cóntrary tó the trend shówn by Fígúre 6.12b fór average annúal flóór temperatúre óver the cómplete óf the Nórthern Hemísphere, and has been ascríbed tó the resúlts óf atmóspheríc aerósóls released by úsíng póllútíón.

Thís íncreased díroct retúrn óf sólar strength tó space has been dúbbed 'glóbal dímmíng' dúe tó a lówer ín sólar radíatíón achíevíng elements óf the Earth's flóór, a trend determíned tíll the early Nínetíes. Thereafter, grówíng cóntróls ón emíssíóns fróm Eúrópean and Nórth Ameríican strength statíóns have caúsed a 'bríghteníng-úp' óf the skíes — an grówth ín ínsólatíón. Írónícally, decreased partícúlate póllútíón ís próbable tó exacerbate the cóntempórary glóbal warmíng fashíón, sínce ít had ín element been óffset vía ínternatíónal dímmíng.

An crítícal addítíónal óútcóme óf the cómbústíón óf fóssíl fúels ís the ímpact the góóds óf súch búrníng have ón aír excellent. Ín December 1952, a chílly fóg (smóg) that cóntaíned excessíve tíers óf smóke and súlphúr díóxíde húng óver Lóndón fór nearly a week. Ít changed íntó óne óf the wórst ín a seqúence óf 'pea-sóúper' fógs that descended ón Lóndón at that tíme, and túrned íntó wíthóút delay chargeable fór abóút 4000 deaths thrú brónchíal ínfectíóns and heart attacks. A públíc óútcry caúsed smóke cóntróls and wíder úse óf smókeless stable gasólíne. At the tíme, medícal dóctórs blamed high smoke levels in the fog, but greater latest studies shows that the formation of fairly acidic debris may also had been important. The 1952 London smog had a pH of one.6 — greater acidic than lemon juice.

The introduction of North Sea gas and the siting of strength stations out of doors centres of population dramatically decreased the occurrence of smoke-encumbered fogs. However, they have been replaced by means of a unique shape of air pollutants — NOx and smoke particles from automobile emissions. A thick haze that constructed up from London visitors fumes in the course of 4 windless days in December 1991 without delay induced the deaths of 160 people. Two pollution were notably concentrated inside the London air: NO_2 stages reached 423 elements according to billion, the best degree ever recorded inside the UK, and black smoke particles reached 228 μ g m-three.

Particularly on sunny summer days, the primary respiratory irritant in 'modern' polluted air is ozone (O_3), best certainly plentiful inside the stratosphere. This ground-stage ozone is fashioned from photochemical reactions regarding each NOx and strains of hydrocarbons within the air. Whereas the London smogs of the 1950s have been the result in particular of home coal combustion, the NOx that forms one of the primary ozone precursors comes particularly from vehicle exhausts and energy station flues. While the consequences of ground-level ozone can have intense effects for human fitness, the results of this pollutants may be even further attaining — natural ecosystems go through and agricultural production falls. Already, rice yields in polluted parts of Asia are idea to be 10% lower than they could were below 'ozone-smooth' situations, whilst soy bean vegetation are even more inclined, with a projected 30% drop in yield by 2020.

3.Four Pollution answers?

At the 1992 Earth Summit in Rio de Janeiro, it became agreed that the important industrialised international locations might stabilise their CO_2 emissions from fossil fuels at 1990 degrees by the year 2000. This decision became taken frequently to alleviate viable international warming caused by the greenhouse impact, however this type of policy might additionally stabilise emissions of other principal members to atmospheric pollution and acid rain, which include oxides of sulphur and nitrogen. Since then, the Kyoto Protocol of 1997 committed signatory countries (but now not america, Australia, India and China, which did no longer signal) to reducing annual greenhouse emissions to five% below 1990 stages by means of 2012. In terms óf acídíficatíón and aír fírst-class, the 1999 Góthenbúrg Prótócól fór the abatement óf acídíficatíón, eútróphícatíón (water póllútíón resúltíng fróm ímmóderate plant nútríents, cónsístíng óf NÓx) and gróúnd-stage ózóne calls fór óf íts sígnatóríes the fóllówíng díscóúnts (relatíve tó 1990 tíers) by way óf 2025: súlphúr emíssíóns thróúgh as a mínímúm 63%; NÓx emíssíóns wíth the aíd óf fórty óne%; únstable órganíc carbón emíssíóns thróúgh 40%; and ammónía emíssíóns by means óf 17%.

Ín Sectíón 2 and ín Fígúre 8 (C eventúalítíes) yóú saw the dramatíc capacíty ímpact óf changíng fóssíl fúels wíth the aíd óf óppórtúníty renewable and núclear resóúrces, wíth a fórce tóward extra green útílízatíón óf ówer. Yóú múst alsó have mentíóned the caveats cóncerníng the hígh pace óf grówíng and deplóyíng súítable technólógíes tó gaín that. An óppórtúníty víew ís that húmaníty have tó lessen íts general strength cónsúmptíón hastíly and sígníficantly. That appróach all óf ús cóúld óught tó úndertake óppórtúníty ways óf dóíng matters: greater pówer cónservatíón at hóme and at wórk; less tóúr órdínary; hígher úse óf mass transpórt; úníqúe lífe and expectatíóns, and só fórth.

Renewable energy assets múst be evólved íf we are tó wean óúrselves óff fóssíl fúels. Wínd electrícíty ís develópíng wíthín the ÚK and ín Eúrópe — bút nót wíthóút sóme cóntróversy óver vísúal ímpact ón the landscape — and sólar technólógíes are enhancíng ín terms óf perfórmance and fee. Núclear energy óught tó meet the demand fór strength even as these new technólógy matúre, wíthóút CO_2 beíng pródúced.

Anóther techníqúe, hówever, may alsó óffer a part óf the answer tó decreasíng CO_2 emíssíóns fróm pówer statíóns — CO_2 seqúestratíón. Thís experímental era seeks tó take away CO_2 at sóúrce, í.E. Fróm wíthín a electrícíty statíón, befóre ít has a chance tó break óút íntó the súrróúndíngs (Fígúre síxteen). Ín óther wórds, the CO_2 ís wíthóút delay transferred tó lengthy-tíme períód stórage, ín preference tó beíng bróúght tó the carbón cycle. Theóretícally, súch seqúestratíón may want tó exceed herbal príces óf carbón búríal. Próbably the símplest answer ís tó púmp CO_2 íntó póróús hówever sealed-óff rócks that after held óíl ór fúel. Ít ís feasíble that sóme CO_2 may be prómpted tó fórm carbónate mínerals ín the hóst róck, í.E. A stróng and cónseqúently móre stróng shape óf garage. Anóther póssíbílíty ís explóítíng the líqúefactíón óf CO_2 at excessíve pressúres and ínjectíng líqúíd CO_2 íntó deep ócean basíns. Líke methane, CO_2 alsó can shape a stable fúel hydrate ín sea-flóór sedíments; a símílarly póssíbílíty fór seqúestratíón. There are óf cóúrse enórmóús technícal demandíng sítúatíóns that need tó be resólved. Súch technólógícal 'fíxes' wóúld alsó regíón a heavy mónetary búrden ón cónventíónal CÓ2 emíttíng electrícity statíóns, makíng ímplementatíón nót líkely ín rapídly grówíng regíóns cónsístíng óf Chína and Índía. Móreóver, thís techníque dóes nóthíng tó relíeve CÓ2 emítted óútsíde strength statíóns, partícúlarly fróm grówíng avenúe and aír shíppíng (nów apprecíably móre than fróm energy generatíón).

Fígúre 16 Díagram dísplayíng varíóús alternatíves fór the seqúestratíón óf carbón díóxíde.

Nó technólógy are cúrrently envísaged that wóúld selectívely extract CÓ2 fróm aír ín the vólúmes vítal tó make any appreciable dístínctíón tó íts óverall cómpósítíón. There ís óne easíer means, hówever, phótósynthesís ís únfastened and efficíent, bút desíres encóúragement. Plantíng súfficíent búshes tó cówl a place the síze óf Aústralía ís thóúght tó have the pótentíal tó lessen atmóspheríc CÓ2 tó tíers cómpatíble wíth a stróng weather. Óther órganíc strategíes óf seqúestratíón cónsíst óf 'fertílísíng' ópen óceans tó ínspíre phytóplanktón bóóm, whóse dyíng, sínkíng and búríal ón the ócean flóór míght repaír carbón ín lengthy-term stórage, albeít at the feasíble rate óf próvókíng ecósystems.

Fóúr Próspects and óppórtúnítíes fór the sectór's energy destíny

Befóre specúlatíng ón what míght líe ahead fór the glóbal energy scene, ít may assíst tó check the majór traíts óf the varíety óf electrícíty sóúrces avaílable.

Fóúr.1 Cúrrently tó be had assets

Fóssíl fúels are the essentíal wórldwíde pówer súpply at the start óf the twenty fírst centúry, bút ón cómbústíón they all emít CÓ2, whích cóntríbútes tó glóbal warmíng, and varyíng amóúnts óf SÓ2, ín cónjúnctíón wíth nítrógen óxídes, whích all mótíve acíd raín. Fúrthermóre, they are a fíníte úsefúl resóúrce.

Cóal ís the móst cónsíderable óf the fóssíl fúels hówever has the bóttóm 'strength densíty' and the greatest póllútíón abílíty. Cómmercíal cóal íncórpórates fíve-10% ór greater óf nón-flamable míneral ímpúríty (ash), aróúnd 1% óf súlphúr, and dífferent elements ín hínt qúantítíes, ínclúsíve óf chlóríne, úraníúm and arseníc. Só-knówn as smóóth cóal technólógy ímpróve cómbústíón perfórmance and decrease póllútants (see Sectíón 4.2). As we saw ín Bóx 1, hówever, the ÚK Góvernment móved faraway fróm cóal thróúghóút the late 1980s and early Níneteen Nínetíes, nótwíthstandíng the trúth that cóal stays the ÚK's móst cónsíderable energy úncóóked fabríc. Ón a glóbal scale, cóal represented aróúnd 22% óf prímary electrícíty íntake ín

2002 (Fígúre 2), and there's súffícíent cóal tó maíntaín call fór at present degrees fór a cóúple óf centúríes, even wíthóút any new reserves beíng fóúnd.

Óíl ís a móre flexíble gas than cóal, wíth a better 'electrícíty densíty' and lów tó 0 ash cóntent materíal. Ít stíll pródúces súlphúr and nítrógen óxíde emíssíóns ín addítíón tó CÓ2 (even thóúgh less than cóal ín step wíth únít óf energy acqúíred). Mótór aútómóbíle exhaúst gases can caúse excessíve tíers óf póllútíón ín cíty areas, even íf decreased thróúgh catalytíc exhaúst strúctúres. Glóbal óíl reserves are súffícíent fór several decades at módern (2005) íntake stages, even thóúgh lócatíng and explóítíng essentíal new reserves may be an íncreasíng númber óf tóúgh and híghly-prícéd.

Gas has the híghest 'pówer densíty' óf all three fóssíl fúels and ís extensívely held tó be a 'easy' fúel, becaúse ít leaves nó ash and íncórpórates wíthóút a dóúbt nó súlphúr; bút búrníng gas nevertheless pródúces nítrógen óxídes as well as CÓ2 (even thóúgh less than cóal ór óíl ín keepíng wíth únít óf strength óbtaíned). Glóbal reserves óf herbal gas alsó are enóúgh fór númeróús decades at módern (2005) levels óf call fór.

Núclear físsíón ís wíth the aíd óf a lóng way the maxímúm fócúsed electrícíty súpply cúrrently avaílable, and the real era óf núclear strength releases nó CÓ2, SÓ2, nítrógen óxídes ór óther chemícals. Hówever, úntíl very stríctly cóntrólled, radíóactíve vía-merchandíse can cóntamínate aír, water and sóíl óver extensíve regíóns fór centúríes tó míllennía. Knówn wórldwíde reserves óf úraníúm are enóúgh tó últímate the lífe óf reactórs tó thís póínt cónstrúcted, and nícely past.

Bíómass, maínly ínsíde the shape óf fúel wóóds and charcóal, represented róúnd 10% óf wórld númber óne energy cónsúmptíón ín 2002, ín large part wíthín the develópíng wórld (Fígúre 2). Bíómass ís renewable íf ít's far ate úp ón the same fee as new vegetatíón are grówn, wheréín case the CÓ2 released all thróúgh cómbústíón cóúld be óffset thrú úptake by úsíng develópíng plants. Bíófúels may alsó be deríved fróm wastes, lóts óf that are órganíc ín startíng place (e.G. Dríed yak dúng ín Tíbet, súgar cane waste and ríce húlls ín Índía).

Alternatíve electrícíty resóúrces are all 'clean' ín addítíón tó renewable; the 'gasólíne' charges nóthíng and cóntríbútes neíther tó ínternatíónal warmíng nór tó acíd raín. Theír drawbacks cónsíst óf lów strength densíty só that very húge ínstallatíóns are wanted fór góód sízed energy technólógy, erratíc geógraphíc dístríbútíón, íntermíttency óf delíver, and wíthín the maín they're nót transpórtable, só strength statíóns need tó be búílt clóse tó the pówer súpply. There alsó are envírónmental príces óf búíldíng the plant (e.G. The materíals needed fór wínd generatórs). Nevertheless, theír blended pótentíal ís large, and plenty óf are adaptable tó nearby strength technólógy.

4.2 Developíng cúrrent pówer systems

fóúr.2.1 Cleaner fóssíl fúels

A style óf strategíes have been advanced fór pródúctíón líqúíd ór gaseóús hydrócarbón fúels fróm cóal ór crúde óíl (ín sóme cases tó redress a defícít óf a specífíc gasólíne ín a únmarríed úníted states). An ancíent ínstance ís 'tówn fúel', ín partícúlar hydrógen and carbón mónóxíde wíth mínór amóúnts óf methane, that ís pródúced by adverse dístíllatíón óf cóal wíthín the absence óf aír (pyrólysís) and wíth steam ínjectíón. The prócedúre becóme assócíated wíth the manúfactúríng óf cóke, cóal tar and óther índústríally benefícíal materíals. Ín the ÚK, cíty gas túrned íntó a maín dómestíc and índústríal gas fróm 1804 tó the 1970s, whíle ít changed íntó hastíly changed by úsíng natúral gas fróm the Nórth Sea. Ít became addítíónally cómprísed óf óíl ín the ÚK amóng 1960 and 1975.

Plentífúl resóúrces óf óíl and gasólíne have cúrbed stúdíes íntó grówíng thóse technólógy; ín any case, why tróúble wíth cómplex cónversíóns when the stúff ín realíty cómes óút óf the flóór genúínely ready tó apply? Hówever, the bíg reserves óf cóal may addítíónally óffer a sóúrce óf óíl wíth the aíd óf líqúefactíón, as well as greater sóphístícated gasífícatíón. Pródúctíón óf óíl fróm cóal becóme fírst advanced ón a cómmercíal scale ín Germany thróúghóút the Thírtíes and 40s, and has óperated ín Sóúth Afríca becaúse the míd-1950s (Fígúre 17), tríggered tó begín wíth thróúgh the úsa's lóss óf óíl, cónsíderable cóal depósíts, and móre and móre ísólated pólítícal stance. Althóúgh 3 plants cóntínúe tó be óperatíónal, the real fees óf pródúctíón aren't cómpetítíve wíth crúde óíl, úntíl óíl expenses retaín tó úpward púsh. Althóúgh the Sóúth Afrícan vegetatíón appóínt a twó-stage system (gasífícatíón óbserved thróúgh catalytíc synthesís), stúdíes presently favóúrs the ímprévement óf a óne óf a kínd, extraórdínaríly lów-temperatúre, catalytíc manner whereín the cóal ís fírst díssólved ín a súítable sólvent earlíer than treatment wíth hydrógen.

The Sasól plant ín Sóúth Afríca that cónverts cóal íntó óíl merchandíse.

Óther nón-tradítíónal resóúrces óf líqúíd petróleúm are antícípated tó symbólíze a fúll-síze, úntapped úsefúl resóúrce (they ínclúde tar sands, óíl shales and heavy óíl. Hówever, these are nót easy fúels, and ín the essentíal they reqúíre wídespread (and lúxúríóús) prócessíng tó súpply petróleúm akín tó crúde óíl. The íssúe ín theír explóítatíón, and theír pótentíal envírónmental effects, have retarded theír ímpróvement thús far, and may maíntaín tó accómplísh that únless the wórld stays hóóked ón petróleúm as an pówer sóúrce past the lífetíme óf cónventíónal reserves. Methane hydrates, cúrrently resídíng ín seaflóór sedíments and a few terrestríal settíngs, cónstítúte massíve capabílíty sóúrces. They alsó present a chíef abílíty chance tó weather, múst warmíng óf deep water destabílíse them. Yet theír fórmatíón entaíls every óther gasólíne hydrate that íncórpórates carbón díóxíde, and methane-hydrate explóítatíón cóúld be cómbíned wíth sea-gróúnd CO_2 seqúestratíón.

Fóúr.2.2 Renewable resóúrces óf energy

Thís sectíón fócúses ón a few wellknówn próblems regardíng the destíny explóítatíón óf renewable sóúrces óf pówer. Ít appears ínevítable that these exceedíngly easy pówer sóúrces wíll see vast develópment all thróúgh the 21st centúry. Ónly hydrópówer has been explóíted near abílíty thús far. Ínvestment and envírónmental dísrúptíón remaín óbstacles fór fúrther hydró schemes, each ón a húge scale ín grówíng cóúntríes and ón a small scale anywhere. Thís ís aúthentíc fór sóme óf the renewable energy sóúrces, althóúgh sóme natíóns have fóstered maín renewables índústríes (e.G. Wínd energy, óffshóre ín Denmark and ónshóre ín Germany).

Írónícally, the geógraphíc dístríbútíón óf renewable electrícíty pótentíal ís júst as chóppy as that óf fóssíl and núclear fúels, só that a few natíóns have a múch móre endówment than óthers. The ÚK, fór ínstance, ís próperly-lócated fór wínd, wave and tídal strength, bút a lót less só fór hydró, sún and geóthermal assets. Ín fútúre súch glóbal íneqúalítíes ín electrícíty sóúrces wíll múst be smóóthed óút (as these days) vía glóbal trade, próbably óf generated energy.

Anóther massíve íssúe wíth many renewable resóúrces ís íntermíttency óf delíver míxed wíth únpredíctabílíty; the realíty that the qúantíty óf electrícíty generated regúlarly varíes ón shórt tímescales. Sóme sóúrces are únpredíctable (wínd and wave energy), óthers are cyclícal (tídal, sún), and óthers are extra cónsístent (geóthermal). Ín desígníng an strength delíver gadget that may gúarantee tó meet call fór, a delícate balancíng act míght be reqúíred. Ít ís nót líkely that renewable sóúrces ón my ówn óúght tó óffer a stable energy delíver, partícúlarly as maxímúm ínternatíónal lócatíóns wíll nó lónger have the whóle varíety óf renewable óptíóns. Thús, óne óf the móst vítal capabílítíes óf destíny strength súpply systems wíll be íntegratíón, thís ís, the capabílíty tó cóntról a cómplex míx óf dífferent pówer assets as predícted fór 2025 ín Fígúre 18 tó pródúce a stable pówer delíver whích can sóng the call fór cúrve.

A predíctíón óf the qúantíty and períód óf úse óf varíóús types óf generatíng plant ín the ÚK óver a year (8760 hóúrs) róúnd 2025, and the órder ín whích they míght be úsed as the pówer demand róse fróm the nón-stóp base lóad — ón the left. Readíng the graph fróm left tó ríght, as extra electrícíty ís needed that ís próvíded fróm a blend óf cóal and gasólíne pówer statíóns, and greater únpredíctable and íntermíttent renewable resóúrces. Thís determíne makes úse óf cónstrúctíve prójectíóns fór the develópment óf renewable technólógíes, and assúmes cónsístent fúndíng ínsíde the núclear índústry. 'Waste' represents strength generated fróm varíóús waste merchandíse (e.G. Sewage slúdge, landfíll gas, íncíneratíón óf múnícípal garbage).

Sóúrces whích are desígned tó súpply strength cónstantly (ín partícúlar núclear, tídal, waste, bíómass and geóthermal) míght be harnessed tó óffer the base lóad, í.E. The mínímúm cóntínúóús call fór fór electrícíty between appróxímately 3.00 and síx.00 AM ín the cóúrse óf súmmer níghts. Óther resóúrces can be bróúght ón-círcúlatíón tó address íntervals óf extended call fór, whích varíes dúríng the yr and alsó ín the cóúrse óf each day. The móst demand, ór tóp lóad, takes place best ín bríef róúnd 18.00 PM ón wínter evenings, fór abóút 300 hóúrs ín a year. The qúíck, múltíplíed demand then may be happy thróúgh dependable, fast-reactíón energy statíóns the úsage óf óíl ór hydrópówer fróm púmped stórage schemes. Demand at sóme stage ín daylíght hóúrs hóúrs, wínter and súmmer seasón, ís specífícally happy thróúgh pówer statíóns that búrn cóal and natúral fúel, súpplemented by means óf renewable sóúrces that are less predíctable (wínd and wave) ór whích vary thróúghóút the yr (sún and hydrópówer). Althóúgh cóal and natúral gas stíll dómínate the schematíc energy management scheme shówn ín Fígúre 18, theír typícal úse ís decreased fróm that ín the early 21st centúry thróúgh the deplóyment óf óppórtúníty pówer resóúrces.

Research at Óxfórd's Envírónmental Change Ínstítúte índícates that íf íntermíttent renewable assets cóntríbúted a enórmóús própórtíón óf pówer (>20%) — and they may óúght tó fór a sústaínable destíny wíth mínímal ínternatíónal warmíng (Fígúres níne and 12) — enórmóús backúp era pótentíal cóúld be wanted. Hístórícally, thís has been wíthín the fórm óf fóssíl-gasólíne energy statíóns, whích ínclúdes the maxímúm hígh-prícéd 'spínníng reserve', whereín túrbínes are certaínly rótated wíthóút generatíng hówever ready tó súpply energy ríght away. Varíóús measúres, bút, can lessen the want fór thís fóssíl-gasólíne backúp:

dístríbútíng ínstallatíóns (e.G. Wínd túrbínes) wídely, ón the basís that the wínd ís úsúally blówíng sómeplace

the úse óf a varíety óf dífferent íntermíttent strength resóúrces, partícúlarly thóse that are partíally cómplementary (e.G. Súnny weather regúlarly methód míld wínds, and více versa)

matchíng pówer resóúrces tó períóds óf hígh demand (e.G. Wíntry weather call fór ís matched thróúgh elevated wínd and wave pótentíal)

replacíng fóssíl-fúel backúp strength statíóns wíth dífferent stórable electrícíty assets.

Thís latter factór híghlíghts the remaíníng predómínant díffícúlty wíth renewables — garage. Whereas fóssíl fúels are wíthóút díffícúlty transpórtable stórehóúses óf electrícíty, renewable assets shóúld be harnessed and the energy úsed straíght away. Íf sóme óf the súrplús electrícíty fróm renewables pródúced at ínstances óf lów call fór (e.G. Sún pówer wíthín the súmmer, ór níght tídes) may be stóred geared úp fór release whílst call fór róse, móst óf the próblems óf renewable súpply can be sólved. The Dínórwíc púmped stórage scheme ís the best ínstance óf óne óf these garage scheme ínsíde the ÚK at the tíme óf wrítíng (2005), hówever there's abílíty fór cómparable schemes ín the shape óf tídal reservóírs. Óf path, súch geógraphícally fíxed appróach óf stóríng súrplús energy are maxímúm súíted tó lócal and at móst regíónal pówer planníng. Anóther feasíble answer dóes óffer a way óf transpórtíng stóred strength. Ít ínvólves electrólysís óf water tó shape extra effórtlessly pórtable hydrógen and óxygen, and bóth úsíng fúel cells tó recómbíne them, releasíng electrícal electrícíty, ór búrníng hydrógen as a gasólíne. These technéqúes shape part óf a fútúrístíc fínancíal machíne based tótally ón hydrógen.

Fóúr.Three Fútúre pówer strúctúres

4.3.1 The hydrógen fínancíal system

The best úse óf hydrógen as a gasólíne depends ón the respónse:

Eqúatíón label:(1)

whích takes place whíle hydrógen ís búrnt ín aír. The reactíón ís partícúlarly exóthermíc, yíeldíng 1.21×108 J kg-1, róúghly twíce the calórífíc cóst óf petróleúm pródúcts (0.Fíve×108 J kg-1 fór petról). Hydrógen ís as a cónseqúence a nón-póllútíng gas wíth a hígh 'energy densíty'. There are símply tróúbles, the densíty óf líqúíd hydrógen ís ten tímes less than that óf petról, and hydrógen bóíls at - 253 °C (20 K), só there are ímpórtant, bút nó lónger ínsólúble, próblems óf transpórt and garage.

Hydrógen cóuld cónsequéntly be úsed as an pówer stórage medíum tó smóóth óut the vagaríes óf únpredíctable flúctúatíons ín strength delíver fróm renewable resóurces. Fígúre 19 íllústrates the stórage ídea; ín thís case wínd pówer ís úsed tó electrólyse water íntó íts thíng gases — the óppósíte óf Equátíón 1 and só an endóthermíc system. The electrícíty úsed ís hence 'stóred' ín the gases and may be released whíle they are recómbíned. Búrníng the hydrógen ís óne way óf dóíng thís; anóther manner ís tó feed hydrógen tó the anóde and óxygen tó the cathóde óf a fúel cellúlar, where they may be recómbíned íntó water. As a end resúlt óf íónísatíón methóds íóns mígrate between the electródes óf the cell and electróns gó wíth the flów ín the external círcúit, pródúcíng a úsable díract electríc módern-day. The ínverter shówn ín Fígúre 19 ís requíred tó transfórm thís íntó alternatíng present day tó be úsed ón an strength delíver gríd.

Stórage óf únpredíctably flúctúatíng pówer resóurces fróm wínd túrbínes (ór fróm sóme óther sóurce) may be achíeved by electrólysís óf water íntó íts thíng gases accómpaníed by theír recómbínatíón ín a gas cell at tímes whíle electríc strength ís needed.

Electrólysís ís símply óne óf several methóds vía whích hydrógen fór gas cells can be pródúced; óther prócedúres íncúde the thermóchemícal splíttíng óf water by means óf hígh-temperatúre chemícal reactíons and the bíóchemícal líberatíón óf hydrógen by a few vegetatíón and algae (thróúghóut phótósynthesís) ór mícró órganísm (by way óf fermentatíón, fór example óf pútrescíble waste). All three strategíes óught tó fórm a part óf a fútúrístíc íncúded system prímaríly based ón hydrógen as the pówer próvíder (Fígúre 20) — the hydrógen fínancíal system. Rather than beíng líquefíed, gaseóus hydrógen may be stóred ín hígh-straín vessels ór as steel hydrídes, ór wíthín the lónger term, póssíbly, ín depleted aqúífers ór óíl wells. The hydrógen ecónómíc system óffers a prómísíng óppórtúníty tó the prevaíling fóssíl gas ecónómy, hówever massíve technícal develópment óf gas cells and bóth stórage and shíppíng óf hydrógen ís requíred befóre óne óf these gadget ís víable. And, óf róute, there shóuld be a decísíve and dramatíc shíft tó grówíng the úsage óf renewables as prímary strength assets.

Elements óf the hydrógen fínancíal system. Energy sóurces are shówn ón the tóp, cónversíón methóds wíthín the center and the garage, dístríbútíón and úse óf hydrógen at the bóttóm.

Fóúr.Three.2 Núclear fúsíón

When very excessíve speed núcleí óf líght atóms cóllíde and cóalesce tó fórm large núcleí — núclear fúsíón, any súrplús mass ís cónverted íntó energy accórdíng tó the Eínsteín relatíón, E = mc2. Úncóntrólled núclear fúsíón ís the ídea óf the hydrógen bómb and pówers the Sún and stars, hówever can ít be cóntrólled and harnessed? Fróm the Fíftíes ónwards, lavíshly fúnded research has púrsúed thís púrpóse, reachíng the fírst cóntrólled laúnch óf fúsíón energy fróm the Jóint Eúrópean Tórús (JET) test ín 1991.

The fúsíón reactíóns whích have been stúdíed are thóse óf deúteríúm and trítíúm (the heavy ísótópes óf hydrógen, cóntaíníng, respectívely, óne and neútróns plús a prótón) whích fúse tó pródúce eíther helíúm-three ór helíúm-4. Fór ínstance:

Equátíón label:(2)

ór, Deúteríúm (hydrógen-2,) cóntaíns zeró.1/2% óf the hydrógen atóms ín natúral waters; the óceans inclúde 4.2×1013 tónnes óf deúteríúm. Ín theóry, thís cóúld pródúce three.4×1012 EJ óf strength íf extracted and fúsed, a íssúe óf 107 extra than the whóle fóssíl fúel bank.

Trítíúm (hydrógen-three, H) dóes nów nót aríse evídently, hówever ís pródúced ín núclear físsíón reactórs that úse water as a móderatór ór cóólant. Ít ís líkewíse a spínóff óf the núclear fúsíón prócess ítself, when líthíúm atóms that shape part óf the cóntaínment system óf the reactór captúre neútróns líberated by means óf fúsíón; trítíúm úse ín núclear fúsíón shóúld cónceívably make bígger thís súpply óf electrícíty nearly índefínítely. Hówever, trítíúm ís radíóactíve wíth a half óf-lífe óf 12 years, and bómbardment óf fúsíón reactór vessels by way óf neútróns and dífferent súbatómíc debrís míght próvíde ríse tó a radíóactíve waste hassle — thóúgh ón a far smaller scale than that related tó físsíón reactórs. The predómínant technícal tróúbles ín maintaíníng fúsíón reactíóns are:

núcleí have tó be delívered tó wíthín 10-15 m óf every óther befóre the stróng núclear appealíng fórces can tríúmph óver electróstatíc repúlsíóns

tó acqúíre the ímpórtant kínetíc energíes, temperatúres óf abóút 108 °C are reqúíred.

The present day layóut fór a própósed Internatíónal Thermónúclear Experímental Reactór (ÍTER) ís fór a excessíve-temperatúre plasma cónfíned vía súspensíón ín an excessíve dóúghnút-fashíóned magnetíc súbject (referred tó as a tórús). Síx cómpaníóns (Chína, the EÚ, Japan, Rússía, Sóúth Kórea and the ÚS) wíll fúnd ÍTER, wíth Indía expressíng róbúst hóbby, bút ín early 2005 the reactór web síte had nevertheless tó be agreed úpón. Fúsíón research has three essentíal góals:

damage-even: whílst tótal óutpút strength = general enter energy. Thís was valídated ón the JET test ín the ÚK ín 1997

búrníng plasma: whereín the plasma ís especíally self-heated by partícle cóllísíóns, wíth líttle óutsíde electrícíty needed

ígnítíón: when the plasma generates só múch pówer that nó enter pówer ís requíred.

A reactór wóúld ín all líkelíhóód need símplest tó óbtaín the fírst óbjectíves tó be cómmercíally víable, and ÍTER ís íntended tó íllústrate búrníng plasma fór the fírst tíme. Hówever, a búsíness prótótype reactór ís nót líkely tó be óperatíónal úntíl at least 2050, assúmíng that the stúdíes ínvestment ís maíntaíned at cóntempórary ranges.

Fíve Managíng strength úse ínsíde the fútúre

Anyóne whó belíeves expónentíal grówth can gó ón fór all tíme ín a fíníte wórld ís eíther a madman ór an ecónómíst.

(Kenneth Bóúldíng, c. 1980)

The fínal years óf the 20th centúry added íncreasíng cóncerns óver úsíng all assets, ínclúsíve óf energy, and the úpward púsh óf internatíónal prójects tó cópe wíth the tróúbles. The 1992 Earth Súmmít at Ríó de Janeíró drew úp a 'sústaínable develópment plan' shówíng hów assets, shíppíng, alternate, bíológícal range, agrícúltúre and físheríes may want tó all be cóntrólled tó maíntaín the best óf lífestyles fór destíny generatíóns. Amóng dífferent recómmendatíóns, the índústríalísed natíóns agreed (ín príncíple) tó stabílíse emíssíón óf carbón díóxíde (fróm fóssíl fúels) at 1990 stages by úsíng the year 2000. (Thís túrned íntó nót perfórmed.) Díscússíóns at Ríó have been accómpaníed by the 1997 Kyótó Prótócól, whích aímed fór fíve% únderneath 1990 CO_2 emíssíón degrees wíth the aíd óf 2012.

Wóúld achíevement óf the Kyótó Prótócól púrsúíts prevent the bóóm ín the cóncentratíón óf atmóspheríc CO_2?

Reveal sólútíón

Althóúgh sóme internatíónal lócatíóns have been relúctant tó decíde tó envírónmental prójects, develópíng númbers óf húman beíngs ín the prósperóús sócíetíes óf Western Eúrópe, Nórth Ameríca and Aústralasía have all started tó 'súppóse glóbally, act lócally', startíng úp and súppórtíng prógrammes óf materíals recyclíng, electrícíty cónservatíón and performance, waste redúctíón, and só fórth. The fínal púrpóse ís fór 'sústaínable ímpróvement' (Sheldón, 2005), thís ís, ímpróvement ínsíde óúr ecológícal means, whích cúrrent húmans abandóned ónce they began cónscíóúsly edítíng theír envírónment tó cónstrúct óúr present day civílisatíón.

Tó pósítíóned ít greater explícítly, sústaínable develópment shóúld súbseqúently cóntaín:

phasíng óút extractíón óf nón-renewable assets

accelerated úse óf renewable assets

recyclíng all manúfactúred súbstances

releasíng all anthrópógeníc wastes at fees cómmensúrate wíth herbal cycles.

Ín the early twenty fírst centúry ínternatíónal, the prímary precedence ís tó decrease fóssíl fúel íntake. Úsíng óppórtúníty, renewable pówer resóúrces wíll assíst, and ín a few ínstances, the úse óf recycled and bíódegradable súbstances — even thóúgh a cómplete pówer aúdít may mónítór that greater energy ís reqúíred fór recyclíng a few pródúcts than fór manúfactúríng them anew fróm raw súbstances. (Móre cómmónly, ít's míles the hígh relatíve fínancíal cóst óf recyclíng that deters súch schemes.) Less eqúívócal ís the gaín óf energy cónservatíón. Thís can take regíón bóth ón the súpply facet ór the demand facet. Demand facet measúres are very númeróús, and can cóntaín prócesses whích míght be eíther technólógícal ór sócíal; we dó nót recóllect them here. Súpply síde measúres cóntaín íncreasíng the effícíency óf electrícíty era and dístríbútíón; as an ínstance, múch less than half the strength wíthín the gas gas fór the móst green ÚK strength statíóns ínsíde the early 2000s ís wíthóút a dóúbt tó be had tó the electrícíty cónsúmer. Múch óf thís únúsed energy takes the fórm óf waste heat, whích cóúld be úsed tó heat búíldíngs, as ín Denmark.

Effícíency has been a súbject at sóme póint óf thís únít, hówever partícúlarly carríed óút tó effícíencíes óf cónversíón, as ín sólar PV energy generatíón. The theóretícal móst perfórmance fór thís prómísíng technólógy ís límíted tó róúnd 30% by means óf physícs, and ís cúrrently appróxímately 15%. Yet perfórmance applíes tó all cómpónents óf húman energy úse, a revealíng ínstance beíng the úsage óf strength tó púmp water; the móst fúndamental want óf a present day sócíety. Say the pówer became generated at a cóal-fíred energy statíón the úsage óf a húndred arbítrary gadgets óf prímary strength. Energy lósses there are róúnd 70%, só símplest 30 úníts ínpút the transmíssíón gríd. Transmíssíón cóúld be very green (nínety óne%), púmp vehícles óperate at róúnd 88%, and púmps themselves at róúnd seventy fíve%. Ónce water ís flówíng thrú all óf the pípelínes and valves tó the cónsúmer, dístríbútíón ís set 47% green ín electrícíty terms, ín part dúe tó cónstríctíóns tó the gó wíth the flów óf a víscóús flúíd, and partíally dúe tó leaks. The net resúlt óf thís chaín óf ínefficíency ís that the púmped water íncórpórates best 9.5 óf the aúthentíc a húndred númber óne pówer devíces. Transpórtatíón cóuld be very a whóle lót wórse. After greater than a centúry óf develópment, vehícle engínes delíver nó extra than 13% óf fúel energy tó the wheels, óf which móre than half óf heats the tyres, avenúe and aír. Bút the efficíency ín terms óf benefícíal paíntíngs, takíng peóple fróm síde tó síde, ís a pathetíc 1%, seeíng that 95% óf the mass transpórted ís the aútómóbíle ítself! Móre ór múch less the ídentícal happens wíth each appróach óf úsíng energy tó dó úsefúl wórk.

Technólógícal measúres cóntaín enhancíng the perfórmance óf electrícíty úse and effectíveness óf cónservatíón ín a varíety óf appróaches:

redúcíng heat lóss fróm búíldíngs, wíth the aíd óf ímpróvíng ínsúlatíón, wíndów glazíng, and só fórth.

Makíng extra green applíances súch as bóílers, frídges, líghtbúlbs, cómpúters, phótócópíers, púmps, and dífferent búsíness, índústríal ór dómestíc machínes

ímpróvíng the efficíency óf shíppíng cars, and grówíng mótórs that rún ón óppórtúníty fúels, as an ínstance hydrógen ín fúel cells (Fígúre 21) ór bíófúels

ímpróvíng manípúlate strúctúres só electrícíty ís cónsúmed best when needed, and at the lówest efficíent óútpút levels

recyclíng waste warmness pródúced by sóme búsíness strategíes (e.G. Kílns) fór decrease temperatúre packages (e.G. Dryíng raw materíals ór merchandíse)

úsíng múch less materíals (e.G. Thínner metals ín aútómóbíle shells), ór materíals whích are less pówer-ín depth (e.G. Plastíc, ínstead óf steel, aútómóbíle búmpers).

Thís bús ís pówered vía a gasólíne cell walkíng ón hydrógen fúel.

The clósíng póínt ís part óf a múch wíder varíety óf bóth technólógícal and sócíal measúres gróúped úndemeath the term de-materíalísatíón, whích means that the úsage óf múch less fabríc (and as a cónseqúence múch less electrícíty) — eíther ín manúfactúríng ór cónsúmptíón. Óne example cóuld be pródúct packagíng; thís wíll ín lóts óf ínstances be redúced súbstantíally at súpply wíthóut affectíng the great óf the pródúct, hówever ín addítíón, cónsúmers búyíng the pródúct óught tó re-úse factórs óf the packagíng as óppósed tó súrely castíng óff ít. Súch small sócíal changes may seem trívíal, hówever they cóuld effect pówerfúl módífícatíóns ín sócíety. Óne íllústratíón ís the cómparísón amóng the cónsúmeríst, 'thrów-away' attítúdes óf the óverdúe twentíeth centúry, and the resóurceful practícality óf húman beíngs at sóme póint óf Wórld War ÍÍ, whílst nóthíng changed íntó wasted únnecessarily. Ít cóúld be argúed that the ónly dífference amóng that póint óf crísís and nów's that the threat óf strúggle became perceíved as being tóward dómestíc, dúe tó the fact present day sócíety cúshíóns peóple in the evólved internatíonal fróm the affects in theír cónsúmerísm.

Ín addítíón tó de-materíalísatíon, sócíal methóds tó electrícity cónservatíon wóúld alsó cóntaín rearrangíng óúr lífestyles, bóth indívídúally and tógether, tó lessen the energy reqúíred fór a specífíc carríer. The creatór ís fórtúnate súffícíent tó líve ínsíde ón fóót dístance óf a tówn centre, facúltíes, and dífferent servíces, bút many newer cítíes (ínclúdíng Míltón Keynes) are desígned wíth decrease pópúlace densítíes, só that maxímúm jóúrneys are ímpractícal wíthóút the úse óf a aútómóbíle ór bús. As Fígúre 22 súggests, vehícle and van úse wíthín the ÚK sóared fróm 1952 tó 1990, and cóntínúes tó be g:ówíng, partly reflectíng a fashíón óf fewer húmans accórdíng tó vehícle.

Annúal passenger-kílómetres travelled ínsíde the ÚK, 1952-2000, thróúgh delívery móde. Nóte: Aír travel ínfórmatíón refers tó ínternal flíghts best.

Ín 2005, there were few sígns and symptóms that thís trend was reversíng, despíte the íntródúctíón óf súch measúres as bús lanes, cóngestíón chargíng (e.G. Ínner Lóndón) and múltíple-óccúpancy vehícle lanes (e.G. Leeds). The ímplícatíón ís that peóple ín develóped internatíónal lócatíóns whích ínclúde the Úníted Kíngdóm are fíndíng ít hard tó shake óff óúr addíctíón, as a sócíety, tó fóssíl fúels, and the way óf lífe that reasónably-prícéd, plentífúl electrícíty has só far bróúght ús. Móre wórryíngly, hastíly develópíng internatíónal lócatíóns inclúsíve óf Chína and Índía are experíencíng the ídentícal, dramatíc úpward púsh ín vehícle úse and accómpanyíng cíty póllútíón. Ínternatíónal aír travel, nó lónger blanketed in Fígúre 22, ís a búrgeóníng tróúble, wíth aír síte vísítórs, aírpórts and even the síze óf aírlíners all íncreasíng, at the same tíme as cósts ón many róútes hónestly fall, encóúragíng demand. Ín effect, any savíngs ín pówer becaúse óf cónservatíón measúres are cúrrently greater than óffset by wíll increase ín aútómóbíle númbers and úsage, and pólítícíans are lóth tó úpset pótentíal cítízens by way óf cúrbíng púrchaser demand (as an ínstance, vía móre dracóníaan fúel taxes).

A símílarly díffícúlty ís referred tó as the rebóúnd ímpact. Thís ís the tendency fór índívídúals ór córpóratíóns, when they have saved móney by means óf ímpósíng energy savíng measúres, tó spend that 'extra' cash ón extra strength-eating activities, along with imparting higher great services. For instance, a householder who installs better loft insulation should shop on heating bills. However, they may truly warmth the house to a warmer temperature, or for longer durations, and use the same amount of energy as earlier than. Alternatively, they'll splash out on an foreign places holiday with the cash stored from the lower bills, the usage of strength-intensive air journey that offsets any strength financial savings inside the domestic. One manner governments can counter the rebound effect is to provide incentives for residents to spend such financial savings in methods which can be electricity-frugal in place of electricity extensive. However, in the end the obligation lies with the character concerned, and how much they proportion winning environmental worries — in truth, how much they surely choice '... to preserve the satisfactory of existence for future generations'.

Activity four

The UK is lucky in gaining access to a wide variety of power resources, and the severa options for destiny power supply has fostered tremendous debate on what the pleasant 'strength mix' have to be. Printed underneath are brief, edited extracts from various sources that highlight distinct elements of the energy debate.

Since 1990, the call for for power in the UK has grown with the aid of 25%. Gas power stations now generate the most important proportion of strength within the UK, and the proportion of fuel technology is anticipated to boom similarly to 50% in 2012, following the closure of older nuclear stations and coal stations. At the cutting-edge time, NGT [National Grid Transco] is not anticipating any new nuclear plant life. The UK will quickly emerge as a net importer of fuel, and in 2013/14, sixty six% of gasoline necessities can be imported. Of the current 22 GW gas-fired era, most effective 6 GW has a again-up gasoline option should the supply of fuel be halted. It seems to me that an interruption to the gasoline supply should result in substantial shortfall of generation.

(Simon Griew, National Grid Transco, 2004)

A critical situation faces this united states of america if we do not deal with the looming electricity disaster. By 2020, this united states could be ninety% depending on gasoline for its strength needs and 70% of so as to must be imported from politically risky areas of the sector together with Russia, Ukraine, Iran and Algeria, through pipelines extensive open to terrorist assault. We can be a internet importer of power and all the time sat ón thóúsands and thóúsands óf tónnes óf the cóal reserves wíth whích thís natíón ís blessed. Óúr balance óf bílls wíll gó thróúgh ímmedíately cónseqúently. Clean cóal technólógy are nów aváilable and are beíng fúrther evólved and úsed nót ónly wíthín the ÚS, bút ín Aústralía, Índía, Chína and many dífferent cóal generatíng internatíónal lócatíóns. Ín the Úníted Kíngdóm we píóneered the research íntó smóóth-cóal era, whích túrned íntó deserted by úsíng the fínal Tóry góvernment ín íts haste tó bútcher the míníng indústry that served, and nónetheless shóúld serve, thís natíón próperly.

(Steve Kemp, Natíónal Secretary, Natíónal Úníón óf Mínewórkers, 2005)

AMEC ís óperatíng clósely wíth lóng-term clíents líke BP tó develóp new resóúrces óf súpply ór íncrease the effectíve líves óf cúrrent fíelds aróúnd the arena — freqúently ín adversaríal envírónments. These óíl and gasólíne recúperatíón inítíatíves are móre and móre vítal as the sectór's shares start tó dwíndle ór túrn óút tó be únrelíable — símply óbserve the recent glóbal petról charge paníc sparked vía terrórísm ín Saúdí Arabía.

Nó rely what yóúr víew abóút internatíónal warmíng and the búrníng óf fóssíl fúels, cúrrent actívítíes have shówn hów crúcíal it's míles that we extract as lóts óíl and fúel as ís technólógícally feasíble and só maíntaín óúr carbón-based ecónómíc system feasíble tíll new sóúrces óf strength are develóped.

(Chrís Bónd, Óíl and Gas Technólógy Dírectór, AMEC, 2004)

The ónly renewable súpply presently able tó súpplyíng a gíant amóúnt óf electrícíty ís hydrópówer, and there are few clósíng póssíbílítíes fór large hydrópówer schemes. The ÚK ís aímíng fór 10% renewable technólógy by means óf 2010. Íf thís were tó be fúlfílled by means óf wínd electrícíty, then túrbínes wóúld have tó be ínstalled at a príce óf 40 cónsístent wíth week amóng nów after whích; the cóntempórary príce ís 2 ín step wíth week.

Íf carbón emíssíóns are tó be cóntrólled, núclear pówer wíll múst play an ever grówíng pósítíón ín pówer era. New reactórs prodúce símply 10% óf the núclear waste the antíqúe ónes accóúnt fór. Íf we únlócked óúr cóal ít wóúld remódel the próspects fór the úse óf fóssíl gas, só carbón seqúestratíón ís the key tó the fútúre, cóllectívely wíth the ónes new núclear plants. Í cannót trúst the Úníted Kíngdóm wíll ever get far past generatíng 10% ór só óf íts pówer renewably — and that míght be a heróíc attempt.

(Próféssór Ían Fells, New and Renewable Energy Centre (NAREC), 2004)

Let's all pass núclear, it's the móst effectíve way. Already núclear ís becómíng the persón answer. And clímate alternate ís the núclear lóbby's first-rate weapón: móst effectíve internatiónal warmíng ís extra rísky than bíg prólíferatíon óf núclear strength acróss the wórld.

Malcólm Wícks, the new pówer mínister, rebúts the myths and factóids nów só effectívely spread by the antí-wind-strength fóyer. Nó, túrbínes are nót takíng úp the ú . S . A .: ónly sóme 800 hectares are had tó attaín the ten% target. Nó, they're nów nót únpópúlar: 80% aíd them and síxty síx% wóúld líke a few ín theír vícíníty. Nó, the intermíttent wínd lósíng ís nó hassle, becaúse the farms are únfóld a ways acróss the cóúnty and cúrrent lówer back-úp ís pretty súfficíent. (Eyesóres? The ÚK had 90,000 wíndmílls insíde the seventeenth centúry.) Bút thóse myths are gaíníng gróúnd, alóngsíde the larger myth that nóthíng hówever núclear wíll dó. Hówever, new núclear statíóns wóúld take a decade tó cónstrúct at £2bn every. Só ít's hard tó lóók thís parlíament cómmíssíóníng greater núclear strength.

Everywhere there are ínexperíenced shóóts óf what ís próbably perfórmed, íf severe móney and pólítical attentión were dedícated tó ít nów. Take mícró-technólógy. Yóú can púrchase a small wíndmíll tó stíck insíde the garden ór ón the síde óf yóúr prívate hóme fór júst £900: ít plúgs intó an regúlar thírteen amp hóme plúg, cúts electrícíty bílls by way óf a 3rd and míght feed intó the gríd. Ímagíne íf each adúlt have been gíven a carbón qúóta. Thóse whó want tó fly lóts ór óverheat a bíg hóúse míght have tó búy móre qúótas fróm lów electrícíty cústómers. Ít wóúld have the excítíng aspect ímpact óf redístríbútíng fínances tówards the ónes tóó negatíve tó apply theír strength ratíón.

By blendíng between websítes and cómbíníng technólógíes, yóú cóúld markedly lessen the varíabílíty óf energy próvíded by úsíng renewables. And íf yóú plan the ríght míx, renewable and intermíttent technólógy can even be made tó match realtíme pówer call fór patterns. Thís redúces the need fór backúp, and makes renewables a extreme alternatíve tó standard pówer sóúrces.

Wínd (ónshóre and óffshóre) cóúld realístícally próvíde sóme 35% óf the Úníted Kíngdóm's pówer, maríne and dCHP [dómestíc cómbíned heat and pówer] each 10-15%, and sún cells 5-10%. Ín óther phrases, extra than half óf the ÚK's electrícíty cóúld ín the end deríve fróm intermíttent renewables. The excessíve própórtíón óf wínd ís dúe tó the fact the wínd blóws hardest wíthín the wínter, and insíde the eveníng — whíle demand ís híghest. The dCHP alsó pródúces greater at tóp tímes, when demand fór decent water and heatíng ís alsó móst pówerfúl. Sólar makes a smaller cóntríbútíón, and pródúces nóthíng at níght. Bút ít ís stíll ímpórtant tó have ít wíthín the míx as ít kícks ín whíle wínd and dCHP pródúctíón ís lówest. A maríne-prímaríly based renewable system wórks fírst-rate whíle ít inclúdes bóth tíde and wave. The míxtúre has decrease varíabílíty, ís hígher at assembly call fór patterns, and makes better úse óf hígh-príced transmíssíón ínfrastrúctúre.

(Ólíver Tíckell, qúótíng Graham Sínden, 2005)

Ímagíne yóú cóntról the Úníted Kíngdóm's strength pólícy fór the fóllówíng 50 years. Cónsíderíng the extracts abóve, whích óf the súbseqúent measúres míght yóu take íntó accóúnt tó be essentíal. Bear ín thóúghts that dístínctíve búsínesses míght alsó have dístínct óútlóóks. Óne may alsó fócús ón 'ínexperíenced' ídealísm, sóme óther ón ecónómíc íncentíves. Eqúally, the próblems ís próbably addressed pragmatícally ór ín the cóntext óf that's technólógícally feasíble.

Góvernment's carbón díóxíde díscóúnt targets

retaíníng the núclear chóíce ópen even as develópíng technólógy fór clean cóal and maxímísíng óíl/gas recóvery, and placíng greater emphasís ón carbón seqúestratíón

fúll-síze fúndíng ín research and ímpróvement óf latest and renewable gasólíne assets ón a large scale, cónsístíng óf wínd, wave, tídal, geóthermal, and núclear fúsíón

fócúsíng fúndíng ón íntegratíng óppórtúníty energy sóúrces and enhancíng pówer garage and súpply, ínclúsíve óf the úse óf hydrógen

stímúlatíng greater nearby netwórk and mícró-technólógy schemes, whíle edítíng energy transmíssíón systems tó cómpríse thóse varíóús, wídely dístríbúted mílls ón a índústríal tó hóme scale

deplóyíng legíslatíón, súppórted by way óf mónetary íncentíves and levíes, tó ínspíre pówer effícíency and cónservatíón at all degrees and decrease íntake óf númber óne electrícíty.

Reveal sólútíón

Havíng stúdíed thís path, yóú'll nów recógníse that even thóúgh a remarkable deal ís únderstóód appróxímately hów the type óf avaílable energy sóúrces shape, hów they can be determíned and explóíted, a amazíng deal less ís thóúght abóút the ímplícatíóns óf the úse óf them. Ít wóúld be nó exaggeratíón tó natíón that, at the óútset óf the twenty fírst centúry and líttle móre than twó centúríes becaúse the start óf the Índústríal Revólútíón, húmaníty ís at a decísíve póínt ín íts hístóry. Decísíóns and actíóns, centred partícúlarly ón changíng hów we get the electrícíty that we want, wíll need tó be placed íntó exercíse — faster ín preference tó later.

Húman sóciety's íntake óf númber óne energy has ímpróved hastíly óver the prevíóús cóúple óf húndred years, althóúgh wórldwíde ín keepíng with capíta íntake óf índústríal fúels was kínd óf steady fór the prevíóús cóúple óf a lóng tíme óf the twentíeth century.

The sample óf díscóvery, and fast depletíón, óf fóssíl gasólíne reserves dúríng the 20th centúry, as expected fór ameríca by M. Kíng Húbbert, has prómpted many dístínct órganízatíóns tó try fórecasts óf fútúre electrícity resóúrces and call fór. These fórecasts typícally take the shape óf eventúalítíes that íllústrate hów pówer úsage develóps ín keepíng with varíóús factórs, súch as cóverage chóíces, technólógícal advances, and predíctíóns óf reserves.

The cóal pít clósúres annóúnced by the Úníted Kíngdóm aúthóríties ín late 1992 have been a tradítíónal ínstance óf ways a hígh-príce úsefúl resóúrce can be rendered wórthless óverníght vía ecónómíc and pólítícal elements únrelated tó geólógícal avaílabílíty. Súch abrúpt adjústments make fórecastíng díffícúlt, specífícally wíthín the pówer índústríes.

Fútúre electrícity eventúalítíes generally tend tó cónfórm tó three essentíal kínds: hístórícal íncrease, whích extrapólate past trends; technólógícal restóratíón, whích are expectíng advances ín pówer effícíency and cónservatíón, and zeró grówth, whích endórse that sóciety wíll be persúaded ór fórced tó accept the gíve úp óf mónetary íncrease. Many sítúatíóns íncórpórate elements óf sóme óf these appróaches.

Móst scenaríós cónstrúcted aróúnd the begín óf the 21st centúry are based ón a few degree óf redúctíón ín fóssíl fúel úse, aímíng tó lessen sóciety's relíance ón thóse fíníte, envírónmentally negatíve assets by harnessíng dífferent electrícity sóúrces, whích inclúde núclear and renewables. Ín addítíón, these alternatíves míght alsó allevíate the state óf affaírs where a small varíety óf natíóns maíntaín the búlk óf the sectór's últímate óíl and gasólíne reserves.

The internatíónal dómínance óf fóssíl fúels fór electrícity súpply cóntaíns wíth ít the díscharge óf the 'greenhóúse' gasólíne CO_2 at óne óf these charge that ít accúmúlates wíthín the súrróúndíngs. Thís ís thóúght tó be raísíng glóbal mean súrface temperatúre; internatíónal warmíng has characterízed the dúratíón sínce the Índústríal Revólútíón. Even wíth the móst ecólógically sóúnd transfórmatíón óf the glóbal strength fínancíal system, ímply flóór temperatúre ís predícted tó ríse vía 1 °C wíth the aíd óf the end óf the twenty fírst centúry, and próbable wíth the aíd óf even extra, gíven 'enterpríse as standard' and a grówíng glóbal pópúlace.

Óther emissións, ín partícúlar súlphúr díóxíde, have múltíplíed the acidíty óf raín ín sóme regións, thereby havíng súbstantíal ecológícal effects.

Effects óf 'greenhóúse' warmíng may addítíónally have been ín part hídden by partícúlate emíssións fróm enterpríse, therefóre decreasíng sún energy óbtaíned by a part óf the Earth's flóór. Redúctíón ín súch emíssións wíll bóóst úp ínternatíónal warmíng.

A ÚK góvernment Whíte Paper ín 2003 mentíóned a ímagínatíve and prescíent óf strength delíver ín 2020, wíth a relevant energy gríd fúrníshed thróúgh wave, tídal and wínd energy spónsóred úp thróúgh fóssíl fúel and núclear electrícíty statíóns. Lócal and mícrógeneratíón (e.G. Vía cómmúníty bíomass flówers, phótóvóltaícs, fúel cells, lócal wínd ínstallatíóns) wóúld be extra prómínent, tógether wíth strength cónservatíón measúres, partícúlarly ínsíde the dómestíc regíón. There becóme nó ídea óf vólúntaríly decreasíng cónsúmptíón.

Fútúre ímpróvement óf energy assets may be centered ón several góals: explóítíng ín large part úntapped fóssíl gasólíne assets (e.G. Tar sands); grówíng technólógy fór cleanser fóssíl gasólíne úse; develópíng the explóítatíón óf renewable strength assets; and óvercómíng próblems súch as íntermíttency and energy stórage.

Óne própósed techníqúe tó strength stórage ís tó adapt a system called the hydrógen ecónómíc system, ín whích hydrógen ís úsed as a medíúm tó keep and transpórt the electrícíty generated wíth the aíd óf íntermíttent assets. Fúel cellúlar era míght múst be advanced, bút óne benefít cóúld be that hydrógen ís a dístínctly 'easy' gasólíne.

Anóther capacíty alternatíve electrícíty sóúrce ís núclear fúsíón, whích deríves electrícíty fróm cómbíníng atóms óf hydrógen ísótópes (deút12eríúm and trítíúm) tó shape helíúm. Becaúse thís techníqúe reqúíres a cómpletely excessíve temperatúre plasma, that's cóntaíned ín a róbúst magnetíc díscíplíne, ít has nów nót bút been advanced ón a búsíness scale.

Technólógícal measúres tó lessen fóssíl gasólíne dependence and develóp óppórtúníty energy sóúrces may be súpplemented wíth the aíd óf varíóús technólógícal and sócíal strategíes tó cónserve pówer, whích ínclúde de-materíalísatíon. Hówever, strength cónservatíón can be óffset by úsíng the rebóúnd ímpact, íf fínancíal savíngs are then spent cónsúmíng greater electrícíty.

ᕤᕤᕤ

SIX
CHAPTER-5

Power

With three húndred clean súnny days, óver a dózen perennial rívers and a shóreline óf móre than 7,500 KMs, Índia becaúse the age óf Púranas, had realísed the impórtance óf the sólar and different soúrces óf renewable electrícity and the electrícity they ówn fór the benefít óf íts inhabítants.

Póst-Índependence, Índia's first Príme Miníster, Shrí Jawahar Lal Nehrú while inaúgúrating the Bhakra Nangal Dam (having a abílity tó generate 1500 MW óf Pówer) defíned ít becaúse the 'New Temple óf Resúrgent Índia'. Hówever, except hydró electrícity, the óther twó plentifúl pówer resoúrces - wínd and sún remaíned úntapped in the últimate 70 years particúlarly dúe tó lóss óf pólitical wíll and únviabílity óf applícable technólógíes.

Thís reality ís nót hídden fróm absólútely everyóne that Índia is the arena's foúrth-bíggest carbon emítter with its pópúlace óf óne. Three bíllíon húmans with pówer area cóntríbúting majórly tó the same. Bút inside the cúrrent years, Índia has made húge strídes ín the renewable electrícity area. The Clímate Change próblem acróss the Glóbe has fúrther própelled the Góvernment and Decísión Makers tó bróaden a detaíled blúe prínt fór smóoth and sústainable energy fór all.

As part óf the inítial cómmítments tó the París Clímate Accórd, Índia plans tó redúce íts carbon emíssión depth - emíssión in step with únit óf

GDP - vía 33-35% fróm 2005 levels óver 15 years. Ít ís rúnníng clóser tó pródúcíng 40% óf íts móunted strength capacíty by úsíng 2030 fróm nón-fóssíl fúels. Thís wóuld caúse a fúll-síze shíft fróm cóal-prímarí́ly based electrícíty era tó renewable strength resóurces. Tó acqúíre these díffícúlt facts, ít has tó pródúce óne húndred gígawatt fróm sún, 60 gígawatt fróm wínd, 10 gígawatt fróm bíómass and 5 gígawatt fróm small hydrópówer by úsíng 2022.

And thís seems pretty an úphíll task becaúse the renewable electrícíty devélópment ín Índía ís stíll ín íts nascent level. As ín keepíng wíth the Mínístry óf Pówer, Góvt. Óf Índía, Índía's strength blend ís evólvíng slówly wíth fóssíl fúels assembly 82% óf call fór; Cóal fínal the dómínant gasólíne wíth a 57.9% share óf tótal pródúctíón ín 2018. Hówever, there's alsó a sílver líníng at the back óf the dark clóud, wíth the percentage óf cóal ín the energy míx prójected tó fall tó 50% by 2040, even as the própórtíón óf renewables ríses sígnífícantly. Renewables wíll óvertake gasólíne after whích óíl wíth the aíd óf 2020 as the secónd bíggest sóúrce óf electrícíty manúfactúríng.

As per the Índ Índ Ínternatíónal Energy Agency's (ÍEA) Renewables Repórt, Sólar and Wínd cónstítúte nínety% óf the cóúntry's abílíty íncrease, that's the end resúlt óf aúctíóns fór cóntracts tó develóp electrícíty-technólógy abílíty whích have yíelded a númber óf the wórld's lówest fees fór each technólógíes. The ú . S ., whích presently has lów tradítíónal strength assets ín cómparísón tó the electrícíty wíshes óf the massíve pópúlatíón and the swíftly grówíng fínancial system, can fóster the gíant capacíty óf sún energy. Únder the leadershíp óf Príme Míníster Narendra Módí Índía ís dedícated tówards the devélópment óf renewable energy ínfrastrúctúre. The óne húndred seventy fíve GW target fór 2022 and the fórmatíón óf ÍSA led by way óf Índía and France ís sóme óther example óf the ídentícal. Apart fróm sún, the ús óf a ís alsó explóríng hydró energy pótentíal ín the nórth-eastern states whích can be an hóme tó the hydró electrícíty óppórtúnítíes.

Besídes the abóve, alternate ín the energy míx may even ríde úpón revólútíónary technólógy, devélópíng pówer demand, stróng wínd and sún resóúrces, cóverage assíst, and devélópíng ínvestments et al and wíll make certaín smart, dependable, clean and lów cóst pówer tó óver a thóúsand míllíón peóple wíth an energy íntake bóóm óf fóúr.2% p.A., qúícker than all prímary ecónómíes ín the ínternatíónal, óvertakíng Chína as the bíggest grówth market fór renewable pówer by way óf the past dúe 2020s.

Anóther research by means óf Ún/íversíty óf Technólógy (LÚT) ín Fínland expóúnds that Índía has a bíg abílíty tó transpórt íntó a tótally renewable energy devíce by 2050, as a resúlt óf an abúndance óf renewable resóúrces. Íf símplest we wíll óptímally leverage state-óf-the-art technólógíes tó harness próactíve cóllabóratíón wíth the índústry, academía and electrícíty ínnóvatíón atmósphere, the vícíníty can flów straíght tó lówer príced renewable strúctúres. Súch renewable energy systems can wórks ín partícúlar ón easy energy, sólar strength, wínd pówer and dífferent new age garage answers. Sólar phótóvóltaícs ís the móst ecónómícal electrícíty sóúrce and batteríes satísfy the níght-tíme pówer call fór. Ín addítíón tó maskíng Índía's pówer call fór fór energy, súch gadget símúlatíón alsó can cóver fór seawater desalínatíón and artífícíal herbal gasólíne past dífferent measúres.

With the ríght ínvestments ín súch ínexperíenced technólógíes, Índía ís nícely placed tó óbtaín all thís. Thís ís bíg gíven Índía's búrgeóníng energy call fór and the persístent delíver demand hóle at the síde óf the súmmer shórtages and óútages, the púrsúít ín the díreetíón óf cleaner pówer sóúrces can have a ímpórtant róle ín allówíng the úníted states óf ameríca's transítíón tó a tótally sústaínable electrícíty machíne. Ensúríng thóse ínítíatíves cómfórtable the necessary fínancíng tó allów that ímpróvement, bút, stays a úndertakíng, wíth a massíve percentage óf Sóútheast Asían prójects taken íntó cónsíderatíón únbankable. The bankabílíty óf renewable electrícíty prójects has cónstantly been an próblem ówíng tó óff takers' ínabílíty tó take ín strength and pay fór ít.

Amóngst the númeróús tendencíes that have taken lócatíón ínsíde the sún and wínd strength segments thís year, the ónes that wóúld have an extended-term ímpact ón the pówer qúarter ínclúde bíddíng wíthín the wínd phase, whích cóúld mean that útílítíes míght nó lónger scóút fór wínd web sítes and chóóse wínd túrbíne próvíders thróúgh cómpetítíve measúres. Anóther crítícal strand ís the góvernment cóúld smóóth 20,000 MW óf sún abílíty, whích cóúld perhaps be the bíggest blóck óf pótentíal tó be aúctíóned ín a únmarríed tranche fór the fírst tíme glóbally. The aúthóarítíes's stróng sólve tó heíghtened best reqúírements fór ímpórted sólar phótóvóltaíc (PV) módúles, enfórced vía ínspectíóns wíll ín addítíón help prócúrers get óver 25 years óf módúle exístence. Thís dísplays a natíónal cómmítment tó green energy and índícates hów the úsa ís fast transítíóníng tówards a renewable-targeted fínancíal system expedítíng renewable capabílíty cónstrúct-úp and dóíng away wíth the dífficúltíes beíng encóúntered wíth the aíd óf búílders and manúfactúrers.

The destíny appears bríllíant as nearly 293 glóbal and hóme cómpaníes have dedícated tó generate 266 GW óf sólar, wínd, míní hydel and bíómass-based tótally strength ín Índía óver the next decade. The ínítíatíve míght entaíl an ínvestment óf $310 bíllíón-$350 bíllíón. Fór example, the Ínternatíónal Fínance Córpóratíón, the fúndíng arm óf the Wórld Bank Gróúp, ís planníng tó make ínvestments appróxímately $6 bíllíón vía 2022 ín several sústaínable and renewable electrícíty prógrammes ín Índía.

The Índían energy zóne has an fúndíng capabílíty óf Rs 15 tríllíón óver the súbseqúent fóúr tó 5 years, whích súggests cólóssal póssíbílítíes ín electrícíty era, dístríbútíón, transmíssíón and system. Whíle there may be plenty óf capítal chasíng the póssíbílítíes ín the renewable zóne, there are númeróús dangers that want tó be stóred ín víew, ínclúsíve óf cóúnterparty rísks each ín terms óf develópers and prócúrers.

The próperly ínfórmatíón ís renewable strength garage system marketplace ín Índía ís expected tó wítness stúrdy íncrease, óver the súbseqúent decade, as sóón as the fee óf garage declínes, that ís póssíbly tó happen dúe tó the sheer vólúme íncrease thróúgh the electríc vehícle path. Hówever, the achíevement wíll best be víable whílst the FAME 2 wíll meet íts favóred óbjectíves.

Tó draw a parallel wíth óther natíóns, ín December TESLA's 100MW Hórnsdale Pówer Reserve battery system ín Sóúth Aústralía bróúght a húndred MW íntó the cóúntry wíde pówer gríd ín óne húndred fórty míllísecónds, ímmedíately póweríng 1,70,000 hóúses when the Lóy Yang cóal strength plant all at ónce went ófflíne. Thís testífíes, why electrícíty garage has gró w tó be a cómplementary sólútíón fór renewable pówer, whích ís seasónal and íntermíttent fór makíng súre 24×7, stróng delíver óf electrícíty. The thrúst ón sólar and wínd ínítíatíves has expanded the challenges ín retaíníng devíce stabílíty, that ís encóúragíng búílders tó gúíde electrícíty gríd netwórks wíth battery garage tó assíst manage the varíatíóns ín pówer delíver. Renewable strength tasks súbsídízed wíth battery generatíón shóúld transfórm the strength scenaríó ín Índía. Hówever, the próject ís tó expand a technólógy thís ís súítable fór bíg renewable pówer tasks. As accórdíng tó enterpríse repórts, the deplóyment óf electrícíty garage ís predícted tó gró w óver 40 ín step wíth cent annúally ínsíde the next 10 years, wíth aróúnd eíghty GW óf addítíónal garage pótentíal. We have úndertaken evídence-óf-cóncept ín battery energy stórage strúctúres, whereín bíg líthíúm-íón battery banks are beíng deplóyed ín Delhí.

As Índía's leading renewable energy players wíth a gróss generatión abílíty óf 3,210 MW thrú clean nón-fóssíl sóúrces, we are dedícated tó cónvert the wórld ín sync with the aúthórítíes's ímagínatíve and prescíent óf sellíng renewable energy búíldíng a tótal capacíty óf 20,000 MW wíth the aíd óf 2025, óf whích 30-40 ín keepíng wíth cent míght be based ón nón-fóssíl fúel. The want óf the hóúr ís addressíng the bankabílíty óf renewable pówer tasks whích has úsúally been an íssúe ín Índía, thanks tó óff-takers' lack óf abílíty tó sóak úp electrícíty and pay fór ít.

The electrícíty púrchase agreement strúctúre wíshes tó be reínfórced ín addítíón tó make renewable pówer ínítíatíves greater bankable. There are states whích, thanks tó theír ecónómíc challenges, are nót encóúragíng the have tó-rún fame óf renewables and are fórcíng súch capacítíes tó back dówn whíle wínd velócítíes are destrúctíve. The aúthórítíes, therefóre, shóúld pút ín fórce óúght tó-rún pópúlaríty as an respónsíbílíty fór all clíents tó búy an amazíng percentage óf easy and ínexperíenced electrícíty. We addítíónally want tó address a few challenges cónfrónted wíth the aíd óf energy pródúcers whích cónsíst óf excessíve fúel súpply danger, tíme óverrúns at flóra, and the cónstraíned payíng capabílíty óf the fínancíally weak dístríbútíón útílítíes dúe tó pre-defíned RPÓs óf theír PPAs.

Last bút nów nót the least, íf yóú want tó cóntínúe tó be energy tremendóús and tó make the móst óf renewable strength resóúrces, we can have tó parallelly attentíón ón cómpetítíve merchandísíng óf electrícíty perfórmance practíces as Índía's Energy call fór wíll wítness an expónentíal spúrge próúdly ówníng tó the líghts and cóólíng necessítíes dúe tó the númeróús clímatíc cóndítíóns, the traíts ínsíde the Electríc Móbílíty, grówth óf the Índústríes as well as rúral electrífícatíón. The Wórld Bank ín íts recórd títled 'Útílíty scale DSM óppórtúnítíes and cómmercíal enterpríse fashíóns ín Índía' has pegged Índía's pówer effícíency market at Rs 1.6 lakh /- cróre vía cónsíderíng the qúít úse strength perfórmance póssíbílítíes whích ís 4 tímes the Rs fórty fóúr,000/- cróre ín 2010, tówards the backdróp óf the fúlfíllment óf the Góvernment óf Índía's ÚJALA scheme tó dístríbúte LED búlbs (Bachhat Lamp Yójana). Tíll nów, óver 28 Cróre LEDs had been sóld thróúghóút the cóúntry whích has ended ín electrícíty savíngs tó the músíc óf 36,545 MÚs and prevented tóp demand óf 7317 MW. Ín fínancíal phrases, fínancíal savíngs óf aróúnd Rs. 14,618/- cróres had been achíeved. Thís can even óffer a súperb market fór gróúps manúfactúríng strength green líghts and hóme eqúípment as well as agencíes ófferíng DSM sólútíóns.

Transítíóníng tó a Sústaínable Energy Ecónómy

A sústaínable strength ecónómíc system ís óne that relíably meets demand at reasónable príce and debts fór externalítíes that aren't cóntemplated wíthín the módern valúe óf fóssíl-gas strength (e.G., NRC, 2010a; Tester et al., 2005). Nó únmarríed generatíón, renewable ór ín any óther case, míght be súffícíent tó fúlfíll thóse cóndítíóns ón íts persónal, só we are able tó want a pórtfólíó óf energy óptíóns. Móreóver, new technólógy have tó be incórpórated íntó sócíety, and só, past cónversíón perfórmance and príce per kWh, there are several elements tó take íntó accóunt ín tryíng tó íncrease the própórtíón óf renewable energy ín each internatíónal lócatíóns' era pórtfólíós.

Ín thís bankrúptcy, we stúdy Ú.S.-Chínese cóóperatíón wíthín the cóntext óf integratíng a díffúsíón óf technólógíes ríght íntó a cóhesíve energy system. We may even speak a númber óf the "enablers" óf renewable strength and perceíve barríers tó the prólíferatíón óf renewables tó be able tó múst be tríúmph óver wíthín the medíúm (2020 tó 2035) and lóng term (tó 2050).

MÓVÍNG TÓWARD ÍNTEGRATED SYSTEMS

Alígníng Energy Effícíency and Renewable Energy Góals

Fór the súbseqúent decade, deplóyíng strength effícíency technólógíes wíll be the bóttóm-cóst chóíce fór móderatíng strength demand (NAS/NAE/NRC, 2009a; 2010b), that ís, lóweríng the amóunt óf strength ínpút reqúíred tó delíver an antícípated level óf carríer. Ímpróvements ín strength effícíency wóúld póssíbly even make ít póssíble tó pút óff, ór díspóse óf, the need fór new generatíón ín sóme regíóns (NAS/NAE/NRC, 2010b). Ín the cóntext óf an incórpórated, sústaínable strength financíal system, strength perfórmance can óffset the úsúally hígher charges óf pówer fróm cleanser, móre óften than nót renewable, technólógy technólógy.

Cónsíder as an ínstance, the natíón óf Hawaíí, whích íntends tó lessen pówer útílízatíón by way óf 30 percentage vía 2030 even as próvídíng 40 percent óf the remaíníng technólógy vía renewable sóúrces. Íf módern-day strength úse ís 14,300 GWh, the 2030 íntentíón wóúld be met by means óf decreasíng annúal cónsúmptíón by úsíng fóúr,three húndred GWh, and vía servíng fórty percentage óf the remaíníng lóad (4,000 GWh óf 10,000 GWh general) vía renewable electrícíty generatíón. Alígníng pówer effícíency techníqúes wíth lóng rún renewable energy góals effícacíóúsly wíll íncrease the share óf renewables wíthín the generatíón pórtfólíó. Únless the rísíng call fór fór energy ís addressed, wíll íncrease ín renewables and dífferent smóóth pówer alternatíves can be óffset thróúgh even greater speedy íncreases ín númber óne pówer demand, wíth the stabílíty beíng met by means óf fóssíl fúels.

Chína has pósítíoned strength effícíency at the leadíng edge óf íts rúles tó ímpróve strength secúríty, allevíate stress ón hóme sóúrces (specífícally cóal and water fór thermal strength era), and decrease envírónmental ímpacts as íts fínancíal system expands. Energy effícíency and cónservatíon are actúally a tóp príoríty ín íts energy planníng and índústríal ímpróvement strategíes, as cóntemplated ín íts góals tó redúce strength íntensíty (pówer ate úp ín step wíth únít óf GDP) by 20 percent fróm 2005 wíth the aíd óf the yr 2020. Each próvínce and maín múnícípalíty has been assígned a redúctíón góal startíng fróm 12 tó 30 percent.

Chína has recógnízed that greater effícíent úse óf strength ón the hóúsehóld and emplóyer stages translates íntó ecónómíc savíngs óver the years. Súch savíngs cóúld próvíde a wídespread óffset tó the hígher cóst óf pródúcíng renewable energy (NAS/ NAE/NRC, 2010a). Ín óther phrases, íf strength effícíency technólógíes can captúre príce fínancíal savíngs ín the near term, they cóúld act as a brídge tó the deplóyment óf greater hígh príced renewable electrícíty technólógíes that cóúld ín the end súpplant cónventíónal fóssíl-gasólíne era.

Módernízíng the Gríd

A módernízed gríd ís extensívely cónsídered an crúcíal aspect óf a sústaínable pówer ínfrastrúctúre (see Chapter three fór a technícal díalógúe óf gadgets that cóntaín a módernízed gríd). The exístíng gríds ín each the Úníted States and Chína are nórmally cónsídered ímpedíments tó the múltíplíed deplóyment óf renewables, becaúse ít's far lúxúríóús tó úpgrade them ón the way tó receíve and balance bíg shares óf energy fróm varíable-óútpút assets líke sún and wínd energy. Bóth cóúntríes retaín tó make bíg públíc ínvestments (greater than $7 bíllíón each fór 2010 [Zprýme, 2010]) ín next-technólógy gríd technólógy, and Chína ís spendíng almóst 10 ínstances that amóúnt ($70 bíllíón fróm íts fínancíal recúperatíón búndle) ón new excessíve-vóltage transmíssíón ínfrastrúctúre (Róbíns et al., 2009). Ín addítíón, becaúse a sízeable part óf Chína's strength gríd has bút tó be búílt, certaín regíóns ín Chína óúght tó pótentíally "leapfróg" tó a cúrrent

gríd system and súccessfúlly becóme experímental web sítes that cóúld tell gríd retrófíttíng effórts ín the Úníted States.

A módernízed gríd wóúld have three awesóme blessíngs fór the cómbínatíón óf renewables. Móst vítal, ít wóúld lead tó móre effectíve demand management with the aid óf permítting lóad-transferring ór díspatchable call fór tó clean óút peaks ór take gaín óf óff-tóp wínd era. Secónd, a módernízed gríd shóúld facílítate the prólíferatíón óf dísbúrsed electrícíty technólógy, whích míght enable lócal and ón-web síte era (e.G., even fór a únmarríed cónstrúctíng) based tótally ón easy electrícíty. As díscússed beneath, dísbúrsed generatíón has the advantage óf bearíng ín mínd fast deplóyment óf renewables at the same tíme as mínímízíng the demandíng sítúatíóns related tó the zóníng ór new transmíssíón traces reqúíred tó cómbíne thóse resóúrces dírectly íntó the prevaílíng dístríbútíón machíne. Thírd, a cúrrent gríd wóúld make ít easíer tó cómpríse pówer garage technólógíes and óther íntegratíón servíces íntó the machíne ítself tó assíst óptímíze óverall devíce óverall perfórmance. Útílítíes wíll nó lónger always need tó úplóad stórage fór varíable-óútpút generatíón (e.G., backúp fór each wínd túrbíne) só lóng as there are óther alternatíves ínsíde the devíce tó balance varíabílíty and maíntaín relíabílíty. Cóst-effectíve energy garage cóúld addítíónally permít a applícatíón tó óptímíze tó be had resóúrces and díspatch energy tó córrespónd wíth call fór, ímpróvíng the príce óf móúnted wínd túrbínes and óther varíable-óútpút túrbínes, as well as the fee óf the transmíssíón traces.

The Tehachapí Wínd Resóúrce Area gíves an ínstance óf hów sóme óf these elements míght need tó cóme tógether tó gúíde massíve-scale wínd farms. Cúrrent estímates fór wínd energy ímpróvement ínsíde the Tehachapí regíón general fóúr,500 MW. Róúghly thírteen.Fíve GW óf garage abílíty cóúld be needed tó captúre three hóúrs óf era íf the vícíníty's wínd resóúrces are fúlly develóped, and the wínd farms are wórkíng at cómplete capacíty. Alternatívely, demand míght be díspatched tó apply avaílable wínd. Fínally, as a clósíng mótel, a númber óf the mílls may alsó want tó be cúrtaíled íf trade óptíóns aren't ín regíón tó útílíze the strength whílst ít's far generated.

Dístríbúted Generatíón

A maín advantage óf many renewable electrícíty technólógy technólógy ís that they may be módúlar, thís means that they can be deplóyed at small scales (e.G., ón man ór wóman hómes) and wíthín present dístríbútíón netwórks, fúrníshed that they cónsíst óf apprópríate cóntróls tó hóld vóltage. They are alsó apprópríate fór small, óff-gríd packages. Becaúse móst óf Chína's early experíence wíth the deplóyment óf renewable electrícíty strúctúres has been tó delíver far óff rúral areas, the úníted states has cóme tó be a leader ín small-scale hydrópówer, sún water heatíng, bíógas dígesters, and mícró-túrbínes fór wínd strength cónversíón. Despíte speedy úrbanízatíón ín Chína, the pópúlace cóntínúes tó be nearly 60 percentage rúral, and a great part óf that pópúlace has restrícted access tó pówer. Thús, díspensed era wíll stay a cóncern ín the geógraphícal regíón, and renewable pówer

technólógy wíll enable rúral gróúps tó harness lócally tó be had, easy pówer resóúrces.

Móst cúrrent óff-gríd strúctúres ín Chína are pówered by úsíng a únmarríed úsefúl resóúrce, tógether wíth wínd, and a lót óf thóse systems encómpass electrícíty garage. As Chína búílds and keeps these systems, there may be póssíbílítíes tó (1) búíld hybríd systems that draw ón múltíple resóúrce tó óptímíze electrícíty avaílabílíty, (2) ínclúde stórage cómpetencíes, and (3) expand súítable cóntróls tó maíntaín relíabílíty. Gíven the specífíc attríbútes óf thóse óff-gríd strúctúres (súch as sústaíned natíónal and glóbal ínvestment), they cóúld alsó be premíere tó gríd íntercónnectíóns (whích are restraíned by úsíng wínníng strength rates and víable dísrúptíóns tó the gríd) as próvíng gróúnds fór hybríd systems and stórage technólógy.

Sólar technólógy are góód applícants fór dísbúrsed era. Chína ís the sectór chíef wíthín the manúfactúre and deplóyment óf sólar water warmers, whích are nów óften móre valúe effectíve than fúel water heaters—these technólógy aren't úsed fór generatíón, bút the traíníng ín terms óf íncentívízíng deplóyment at a famíly- ór persón clíent-degree can alsó transfer tó róóftóp PV. Ín ameríca, útílítíes have presented packages (e.G., net meteríng) tó ínspíre hóúsehólds tó ínstall róóftóp PV strúctúres. Recently, útílítíes have wórked dírectly wíth búsíness and índústríal web sítes tó rent róóftóps and ópen areas tó deplóy PV strúctúres; thís facílítates útílítíes ín heat clímates tó satísfy heíght call fór and may póstpóne ór take away the need tó cónstrúct new herbal-gas peakíng vegetatíón. Chína has been a leader ín míxed warmness and pówer (CHP) technólógíes, even thóúgh tó thís póínt thóse have úsúally been cóal- ór fúel-based systems. An area fór destíny stúdíes may be tó develóp renewable-strength-pówered systems whích cóúld óffer heatíng, cóólíng, and electrícíty ón a búíldíng ór cómmúníty scale. Fúel cells are already útílízed ín CHP packages and can úse renewable gasólíne, and sún technólógy are anóther apprópríate candídate fór CHP.

Dístríbúted technólógy can play an vítal cómpónent ín the transítíón tó a sústaínable electrícíty ínfrastrúctúre. Ít óffers blessíngs fór útílítíes, só as tó be capable óf incórpórate new renewable capacíty wíth óut the challenges assócíated wíth zóníng and permíttíng an entírely new web síte fór impróvement. In addítíón, the near próxímíty óf electrícíty túrbínes tó electrícal masses wíll lessen sóme óf the expenses assócíated wíth renewables, súch as transpórtatíón charges and transmíssíón líne pówer lósses. Fór example, ín Chína these days, pówer ís faírly lúxúríóús alóngsíde the cóasts dúe tó the excessíve expenses óf transpórtíng cóal fróm remóte lócatíóns. Fínally, dísbúrsed technólógy shóúld make the electrícal system extra resílíent, whích ís a desírable satísfactóry fór bóth sóftware óperatórs and cústómers. Thís wíll depend ón the precíse era and the lócal dístríbútíón gríd characterístícs. Dístríbúted strúctúres tend tó be extra prícey, ón a per watt fóúndatíón, than prímary statíón ór búlk renewable cómpónents, hówever thís ís dístínctly dependíng ón the present ínfrastrúctúre, retaíl charges fór electríc pówered strength, and óther elements. Últímately, príce effectíveness and relíabílíty cóncerns wíll díctate the deplóyment óf

renewable díspensed technólógy and garage, vís-à-vís fóssíl-fúeled óptíóns ór extensíóns tó present transmíssíón and dístríbútíón netwórks.

Electrícíty-Pówered Transpórtatíón

Bóth ameríca and Chína have shówn an hóbby ín energy-pówered vehícles as a means óf lóweríng dangeróús cellúlar sóúrce emíssíóns, gaíníng a cómpetítíve facet wíthín the grówíng market fór manúfactúríng cars, and lóweríng dependence ón petróleúm. Electrífyíng transpórtatíón strúctúres may alsó lessen a númber óf the vólatílíty assócíated wíth fúel cósts. Althóúgh retaíl energy charges wóúld nónetheless range prímaríly based at the tíme óf day, míxtúre demand, and dífferent factórs, generatíng a bígger própórtíón óf electrícíty lócally óúght tó lessen a númber óf the dangers assócíated wíth dependency ón a cómplícated, wórldwíde cóst chaín fór óíl ímpórts.

Electrícíty-pówered transpórtatíón systems addítíónally have awesóme advantages fór an incórpórated, sústaínable pówer ecónómy. Althóúgh aútómóbíle-tó-gríd garage ísn't víable wíth tóday's electríc mótórs, batteríes, and gríd ínfrastrúctúre, a cómmúníty óf electríc mótórs can (1) act as a cómmúníty óf díspensed chargíng masses that may be túrned ón and rancíd, and (2) thrú próper cómmúnícatíóns strúctúres take advantage óf wínd sóúrces, whích have a tendency tó be móre generíc at níght (whílst many mótórs have tó be móre óptímally rechargíng).

As many stúdíes have próven, dúe tó the fact dísrúptíve technólógíes, whích inclúde renewable strength generatórs, dó nó lónger necessaríly fóllów the same óld evólútíónary dírectíón (e.G., Chrístensen, 1997; NRC, 2009b) they wíll benefít tractíón ín new markets befóre they vírtúally dísplace íncúmbent technólógy. Electrífyíng the transpórtatíón system insíde the Úníted States ór Chína that ínclúdes nón-públíc aútómóbíles (e.G., plúg-ín electríc mótórs [PEVs]), públíc transít, and óther transpórtatíón módes (e.G., electríc pówered-pówered bícycles, whích are already extensívely úsed ín Chínese tówns), cóúld create a pótentíally extensíve marketplace fór pówer era. Whether thís new strength demand wóúld mótíve relíabílíty tróúbles and great valúe íncreases depends ón fee cóntról. Therefóre, ít ís góíng tó be crúcíal tó bróaden príces and applícatíóns that ínspíre aútómóbíle chargíng whílst ít's far móst excellent fór the devíce. Ótherwíse, PEVs cóúld úplóad extra peak lóad, grówíng búrdens ón ínfrastrúctúre and úníversal cósts.

There are severa mónetary and techníca demandíng sítúatíóns tó electrífyíng the exístíng transpórtatíón ínfrastrúctúre ín the Úníted States (NRC, 2010c) and the large and swíftly expandíng transpórtatíón ínfrastrúctúre ín Chína. There alsó are cómpetíng óptíóns tó electrífíed transpórtatíón, cónsístíng óf advanced ínner cómbústíón engínes and hydrógen fúel cells. Thús a assórted pórtfólíó óf transpórtatíón technólógíes can be a múch móre líkely state óf affaírs (NRC, 2008, 2010c) than a whólly electrífíed devíce. An NRC (2010c) óbserve estímates that, wíth the aíd óf 2030, 13 tó fórty míllíón PEVs cóúld be a part óf the Ú.S. Vehícle fleet óf three húndred míllíón and that the cósts and deplóyment óf PEVs wíll rely ín large part ón battery charges

(despíte the fact that chargíng aútómóbíles at níght tó redúce gríd cóngestíón and úse óff-heíght energy era and dífferent cóncerns wóúld cóme tó be móre and móre vítal). Addítíónal factórs tógether wíth góvernment íncentíves, óíl cósts, and envírónmental rúles wíll ín all líkelíhóód affect the deplóyment óf PEVs.

Accórdíng tó Húó et al. (2010), súbstantíal electríc car úse ín Chína, wíthín the absence óf córóllary effórts tó redúce aír póllútants fróm the electrícíty era sectór, óúght tó have úníntended envírónmental affects, even thóúgh electríc pówered cars míght make a cóntríbútíón tó úpgrades ín cíty (í.E., neíghbórhóód and lócal) aír pleasant where vehícle exhaúst fróm ínternal cómbústíón engínes ís nów a prímary póllútant. Nevertheless, the aúthórs endórse that Chína próceed wíth electríficatíón packages ín areas ín whích smóóth, lów-carbón strength sóúrces are tó be had. They alsó própóse that pówer-regíón and transpórtatíón-area pólícíes be coórdínated, despíte the fact that energy-sectór refórm tends tó be slówer than adjústments ínsíde the transpórtatíon zóne dúe tó the cómparatívely lengthy lífespan óf present capítal stóck. Refórmíng thóse sectórs símúltaneóúsly, they argúe, míght hyperlínk the pótentíal benefíts tó húman fítness and the súrróúndíngs.

Úrban Development

Móre than eíghty percent óf the Ú.S. Pópúlace ís úrban, and Ú.S. Cítíes eat appróxímately 75 percentage óf the natíón's pówer and are respónsíble fór a fúrther large share óf greenhóúse fúel (GHG) emíssíóns (Grímm et al., 2008). Chína nów has extra than 500 míllíón úrban resídents, and that varíety ís íncreasíng swíftly. Cítíes are "cóncentratíóns óf hómes and assócíated ínfrastrúctúre, and the cónstrúcted envírónment, a key patrón óf materíals and energy, gíves many póssíbílíties fór savíngs" (WRÍ, 2005). Thús effórts tó cónstrúct a sústaínable pówer fínancíal system can make sízable prógress wíth the aíd óf addressíng the desíres óf tówns.

Althóúgh the resúlts óf tradítíónal electrícíty úse are felt ón a lócal and wórldwíde scale, many óppórtúnítíes tó redúce the ímpact óf energy íntake, ín cómpónent by means óf íncórpóratíng greater renewable strength, wíll be ón the neíghbórhóód degree (NAE/ NRC/CAE/CAS, 2007). Ín addítíón tó grówíng íssúes abóút húman cóntríbútíóns tó clímate trade, tówns óúght tó reply tó íssúes appróxímately aír fírst-rate, rísíng strength cósts, vísítórs cóngestíón, and plenty óf dífferent próblems that can be addressed, as a mínímúm ín cómpónent, vía púrsúíng a greater sústaínable pówer appróach.

Technólógy-based sólútíons cóúld be essentíal fór cónvertíng thís sítúatíón, bút behavíóral adjústments alsó are a príme abílíty sóúrce óf ímpróvement, and tówns can be catalysts fór these changes. Cítíes are already makíng changes thróúgh rúles fór búyíng renewable electrícíty, the really apt úse óf íncentíves and rúles tó ínteract the prívate zóne ín grówíng renewables, and land-úse selectíóns that cóúld effect a tówn's electrícíty prófíle.

Rízhaó, Chína, ís an example óf a tówn whereín a númber óf these factórs—neíghbórhóód and próvíncial aúthórítíes fínancíal aíd fór sún R&D, nearby índústríes availíng themselves óf thóse íncentíves, and pólítícal leadershíp devóted tó deplóyíng

the new technólógy—have cónverged. Thís nórthern Chínese tówn óf 3 míllíón ínhabítants makes úse óf sún technólógy fór almóst all óf íts heatíng (búíldíngs and water) and lóts óf the metrópólís's óútdóór líghtíng fíxtúres

(Baí, 2007). Ín the ÚS, Aústín, Texas, Berkeley, Califórnía, and Madísón, Wiscónsín, addítíónally have very cómpetítíve renewable strength packages, regúlatíóns, and íncentíves that have súbstantíally íncreased renewable electrícíty develópment. The Ú.S. Department óf Energy (DÓE) Sólar Ameríca Cítíes Prógram ís wórkíng wíth many metrópólís góvernments tó expand úrban renewable energy ímpróvement.

Cítíes alsó are well pósítíóned tó edúcate theír cómmúnítíes ón sústaínable pówer úse. Públíc traíníng can búíld aíd fór neíghbórhóód strategíes and stress cóúntry and cóúntry wíde óffícíals tó adópt regúlatíóns that sell sústaínable strength úse. Lócal íníítíatíves tó develóp renewable energy can dísplay what's víable, at what fee, and wíth what exchange-óffs (ÍEA, 2009). Stúdíes have próven that móve-cíty learníng cóúld be very crúcíal fór spreadíng knów-hów abóút grówíng púrífíer strength strúctúres (Campbell, 2009). Thús the systematíc accúmúlatíón and technólógy óf transferable únderstandíng fróm súccessfúl experíments may be extremely effectíve ín transferríng ín the dírectíón óf a sústaínable pówer ecónómy (Baí et al., 2010).

TRANSFÓRMÍNG THE ENERGY SYSTEM

The datíng between technólógy and sócíety, called a sócíótechnícal gadget (Emery and Tríst, 1965), has great ímplícatíóns fór grówíng the presence óf renewable electrícíty technólógy. Althóúgh wídespread prógress has been made ín renewable-energy-related technólógy, research shów that changes ín the strength gadget as a whóle are a "gradúal, paínfúl and extraórdínaríly úncertaín techníqúe" (Jacóbssón and Jóhnsón, 2000). Meaníngfúl transfórmatíón wíll ónly be made whílst technólógy that trade cúrrent practíces are defínítely adópted and tíme-hónóred by úsíng sócíety.

The hígh fee óf renewables (e.G., capítal necessítíes fór technólógy generatíón, the need fór brand new transmíssíón traces, ór the fee cónsístent wíth kWh) ís óften mentíóned as an óbstacle tó theír bóóm and ís regúlarly as cómpared tó the valúe óf cóal-fíred baselóad generatíón. Fór bóth Chína and ameríca, hydrópówer, and móre recently, wínd strength and geóthermal, are the móst ecónómíc renewable strength sóúrces. Ín Chína, bíópówer ís 20 percentage móre cóstly than cóal, and sún strength may be úp as a whóle lót as 10 tímes as expensíve. Ín thís sectíón we ínspect the ínterrelated róles óf góvernments, públíc and persónal stúdíes, and sócíety ín remódelíng pówer strúctúres ínsíde the Úníted States and Chína.

Shapíng a Clean Energy Market

Market mechanisms by myself can nót rewórk the prevaíling pówer system, and technólógical answers are insúfficient except they may be accepted ór integrated intó sóciety. Argúably, a fúndamental úndertakíng fór each the Únited States and Chína ís

Page 158

Súggested Cítatión:"6 Transítióning tó a Sústainable Energy Ecónómy." Natiónal Academy óf Engíneering and Natiónal Research Cóuncíl. 2010. The Pówer óf Renewables: Óppórtúníties and Challenges fór Chína and the Únited States. Washíngtón, DC: The Natiónal Academíes Press. Dóí: 10.17226/12987.×

Add a be aware tó yóur bóókmark

that, becaúse óf beyónd súbsídíes and plentíful hóme reserves óf fóssíl fúels, the general públíc maíntaíns tó expect "cheap" strength, whích púts almóst each múch less-ínstalled technólógy at a dísadvantage (ÍEA, 2010c; Weíss and Bónvíllían, 2009). Chapter 5 defíned óne methód tó addressíng thís, vía mandatíng a selected amóunt óf renewables be cóvered ín the era pórtfólíó. The fóllówíng sectíón detaíls sóme óther, ecónómic system-wide refórms that wóuld ímpact renewables.

Íntervening ín Energy Prícing

The reasón fór aúthóríties interventión ín energy cósts ís that agencíes make chóíces prímaríly based ón the market príce óf energy, whích wón't encómpass the expenses related tó envírónmental harm, clímate exchange, strength secúríty, and dífferent externalíties (NRC, 2010a). As a resúlt, móst órganízatíóns decíde nów nót tó pút ín fórce technólógíes that are sócíally green becaúse, they argúe, the persónal retúrn ís júst tóó lów. The númber óne mechanisms tó adjúst fór thís, ór "tó stage the playíng area" fór easy pówer alternatíves (ínclúding energy effícíency), are dírect pówer taxes, cap-and-change ór cap-and-dívidend prógrams, and centered súbsídíes (ór díscóunts ín súbsídíes fór less apprópriate fórms óf energy). All óf thóse mechanisms have an effect ón the prólíferatión óf renewables wíthín the market, hówever tó varíóus ranges.

Carbón Taxes. Taxes ínvólve placíng a rate sígn and lettíng índústry píck óut the means óf lówering energy íntake. A carbón tax affects úsing device and systems already ín area and presents íncentíves fór the adóptión óf latest technólógy and óperatiónal effícíencíes. Taxes shíp a clean, óbvíóus, cóverage message that the reasón óf the addítiónal fees ís tó perfórm sócíetal góals. The reactión tó súch a tax, hówever, ís únsúre, and empírícal estímates óf elastícíties (the ratíó óf alternate ín rate tó trade ín demand) aren't únique sufficient to are expecting the consequent power financial savings.

A carbon tax might not provide enough incentives for era improvement, in particular given the political problems associated with enforcing a excessive enough tax to provide a substantial incentive. In addition, even though a carbon tax may additionally cause instant financial savings if proprietors of existing flora and gadget reduce their strength consumption, it's going to however impose fees that had been now not expected whilst the investments in era and motors were installed region, raising troubles of fairness. These troubles will be addressed through phasing inside the tax on a preannounced time table.

Cap-and-Trade Systems. As of July 2010, Congress is thinking about enacting a "cap-and-exchange" machine to cap GHG emissions1 at a predetermined degree and

1
HR 2454 changed into exceeded by using the House of Representatives on June 26, 2009. The invoice sets a cap on carbon dioxide emissions that covers approximately eighty five percent of general U.S. Emissions, including emissions from domestic oil refineries.

Issue some of permits equal to that cap. Controlled entities, which include electric utilities and oil refineries, might have to give up a allow for each ton of CO_2 emitted. Because the allows might be traded, an entity could choose to reduce its personal emissions or buy lets in from a permit holder willing to promote, relying on the common overall value. The marketplace fee of permits will be contemplated within the price of production and ultimately exceeded directly to the client. In a few sectors, the permit charge might have the equal effect as an energy tax.

Targeted Subsidies. Both China and america have a precedent—sulfur dioxide (SO_2) pollution—for correcting market screw ups within the strength sector (NAE/NRC/CAE/CAS, 2007). The United States used centered technological solutions (primarily SO_2 scrubbers and fuel switching) to force dramatic discounts in emissions, a pattern China is now following.

Subsidies, either inside the shape of direct charge supports for renewable electricity, or indirect helps through discounts in subsidies for different kinds of electrical era, are every other pricing device. The U.S. Federal government already makes use of subsidies to affect charges. Over the course of 7 years, 2002 to 2009, there have been $seventy two billion in subsidies for fossil fuels and $29 billion for renewables (ELI, 2009). The difference ín valúe ís vítal, bút as became póinted óút ín Chapter fíve, sóme óther vítal cómpónent óf súbsídíes ís theír cónsístency óver the lóng term. Ín thís sítúatíon, that cónsístency, ór lack óf ít, deepens the dívíde between the súbsídíes. Many óf the móst ímpórtant súbsídíes fór fóssíl fúels were wrítten íntó the Ú.S. Tax códe, whíle súbsídíes fór renewables had been súrpassed as bríef ínítíatíves2 (Bezdek and Wendlíng, 2006, 2007). Ít ís líkewíse sízable that appróxímately half óf óf the súbsídíes fór the renewable area had been fór córn-prímaríly based ethanól.

China has a símílar recórds óf súbsídíes fór cóal, strength, and petróleúm. The ímperatíve góvernment regúlates all electrícity fees, and these súbsídíes were úpheld as óblíqúe súppórt fór pówer-ín depth heavy índústríes ín Chína ín addítíon tó a manner tó míld clíent ínflatíon. Ín 2008, a few rate cóntróls had been cózy, these súbsídíes cóntínúe tó be a súbject óf díalógúe and, tó the qúantíty that they preserve fóssíl fúel-deríved pówer cósts artíficíally lów, they'll retaín tó place new renewable energy era at a dísadvantage.

Brínging Clean Energy íntó the Maínstream

There are númeróús cóntempórary examples óf renewable energy technólógy that battle ín the marketplace fór nón-technícal mótíves. Fór example, wínd farm traíts had been behínd schedúle dúe tó aesthetíc wórríes, and waste-tó-pówer facílítíes were antagónístíc ón the ídea óf envírónmental ínjústíce.

2

Fór ínstance, federal tax súbsídíes fór íntangíble drílling príces fór óíl and herbal gas were a permanent fíxtúre óf the Ú.S. Tax códe fór móre than 60 years. Súbsídíes fór renewables, ínclúsíve óf the manúfactúring and fúndíng tax credíts (see alsó Chapter 5 díscússíón) have lapsed and been reínstated several tímes wíthín the past decade.

Hístóríícally, the sítíng and creatíon óf transmíssíón tasks have aróúsed a excellent deal óf públíc and pólítícal cómpetítíon, and the cóntróversy has been reópened wíthín the cóntext óf recent transmíssíón maínly fór renewable energy tasks.

Fór renewables tó reap a sízable share óf basíc energy technólógy, the enterpríse wíll múst penetrate the maínstream strength markets ín each the Úníted States and Chína. Hówever, úntíl very recently, renewable pówer becóme cómmónly referred tó as a gap enterpríse. Renewables wóúld póssíbly attaín maínstream fame by means óf step by step grówíng market percentage. Ín the ínterím, advócacy gróúps, prófessíónal sócíetíes, and enterprise assóciatións ínside the Úníted States and Chína are rúnníng tó accelerate thís trend by means óf cónveníng cómpaníes, dísseminatíng ínfórmatíon, lóbbyíng pólicy makers, and every nów and then úndertakíng R&D (e.G., the Electríc Pówer Research Ínstítúte [EPRÍ] ínsíde the Úníted States).

Ín the ÚSA, every predómínant sóúrce óf renewable strength has a alternate órganísatíon wíth tens óf thóúsands óf members. Ín Chína, the 2 largest índústry ínstítútións are the Chínese Wínd Energy Assóciatíon and íts fígúre órganísatíon, the Chínese Renewable Energy Sócíety. The Amerícan Cóúncíl ón Renewable Energy (ACÓRE), establíshed ín 2001, nów has nearly 1 míllíón payíng índívídúals and these days created a Ú.S.-Chínese applícatíon as íts flagshíp ínternatíónal effórt tó sell díirect hyperlínks between índústry leaders ín bóth natíóns. The effect óf thóse búsínesses ís tóúgh tó qúantífy, hówever theír fast bóóm ís a trademark óf the an íncreasíng númber óf prómínent fúnctíon that renewable energy ís playíng ín tópíics óf ecónómíc devélópment and energy cóverage.

Strengthening Ínnóvatíon

Úníted States

Ínnóvatíon ín renewable electrícíty has cómmonly been related tó strength príces (Weíss and Bónvíllían, 2009). Ín the ÚS, electrícíty R&D ín pópúlar has been greatly prómpted thróúgh the prevaílíng charge óf óíl (and hence the perceíved want fór ínnóvatíon ín strength perfórmance and alternatíve resóúrces). The declíne ín Ú.S. Federal spendíng ón energy R&D has been well dócúmented (e.G., Dóóley, 2008; Kammen and Nemet, 2005; Margólís and Kammen, 1999). Dóóley (2008) nótes that becaúse the míd-Nineteen Níneties, energy R&D has accóúnted fór móst effectíve 1 percentage óf federal R&D expendítúres. Margólís and Kammen (1999) caútíóned that cútbacks, whích began róúnd 1980 fóllówíng the electrícíty dísaster ín the past dúe 1970s, cóúld úndermíne ínnóvatíon pótentíal wíthín the pówer area.

Varíóús ópínións óf federal ínvestments ín easy strength R&D have encóúraged dramatíc wíll íncrease, ón the órder óf $15 tó $30 bíllíón per 12 mónths (Dúderstadt et al., 2009; Kammen and Nemet, 2005; Nemet and Kammen, 2007). As a reference factór, ín 2009, ínspíte óf the óne-tíme ínfúsíón óf ínvestment fróm the Ameríican Recóvery and Reínvestment Act, the Department óf Energy's R&D príce range tótaled abóút $9.5 bíllíón. Móreóver, thís príce range ís break úp amóng prótectíon (~37 percent), basic technólógy (~42 percentage), and strength (~21 percent), with carríed óút strength R&D tótalíng $2.27 bíllíón fór the fínancíal 12 mónths 2010 (AAAS, 2010). Óverall, fúndíng ín renewable strength research has nót been enóúgh tó súppórt massíve deplóyment at súffícíently lów-príce (NRC, 2000; NSB, 2009). Ín sóme índústríes, tógether wíth chemícal cómpóúnds and electrónícs, persónal córpóratíóns fúnd móst R&D—wíthín the Úníted States, federal R&D fúndíng ín thóse sectórs represents ~1 percent and ~0.5 percent, respectívely (NSB, 2010). Prívate córpóratíóns ín these and dífferent índústríes have typícally exhíbíted hígher R&D spendíng/sales ratíós (ëíght-10 percent) than strength útílítíes (~zeró.5 percent). Díect aúthórítíes fúndíng can't make úp fór thís shórtfall, hówever góvernments can óffer leverage wíthóút delay (vía addítíónal ínvestment ín pre-cómmercíal R&D) and nót dírectly (e.G., tax credít fór nón-públíc R&D spendíng).

Públíc and prívate R&D has tended tó emphasíse íncremental úpgrades ín cómmercíalized ór eqúípped-tó-be-cómmercíalized renewable energy technólógy. Góvernment help has alsó tended tó be generatíón-specífíc, that specíalíze ín advancíng wínd generatórs, fór example, alóngsíde a príce/ watt cúrve. Becaúse óf the abúndance óf sún electrícíty, ít has nórmally been cónsídered the maxímúm prómísíng renewable aíd fór brand new dísrúptíve technólógíes (Lewís and Nócera, 2006).

Hówever, fór each generatíón cómmercíally tó be had and prepared fór elevated deplóyment, several óthers ón the draftíng bóard may want tó próbably be "game changers" ín the feel that they míght póínt the way dówn a dramatícally dístínctíve cóúrse tó cóst-effectíve smóóth electrícíty. Althóúgh exístíng technólógíes are predícted tó maíntaín tó enhance and góvernments and prívate enterpríse wíll maíntaín tó pút móney íntó carríed óút stúdíes ín súppórt óf thís, ít's alsó crúcíal that R&D be óríentated tóward lóng-tíme períód góals fór sústaínable strength (NSB, 2009).

Ínnóvatíón ínsíde the Úníted States ís an íncreasíng númber óf beíng mótívated thróúgh úníversíty-índústry partnershíps, whích, ín flíp, generally tend tó emerge fróm and be stímúlated vía góvernment móves (Feller, 2009). The ratíónale fór públíc-prívate partnershíps ín R&D ís tó deal wíth the ínvestment gap fór entrepreneúrs attemptíng tó cómmercíalíze medícal ínnóvatíóns; the caúse fór aúthórítíes help fór thís fórm óf R&D ís that the sócíal charge óf gó back (the benefíts tó sócíety) are móre than nón-públíc charges óf retúrn (fabríc blessíngs tó a selected cómpany) (Shípp and Stanley, 2009). These had been the únderlyíng standards óf Ú.S. Aúthórítíes ínvestments ín renewable strength technólógíes becaúse the Seventíes.

Góvernment ínvestments cóntínúe tó be dírected ín the dírectión óf públíc-persónal partnershíps fór yóu tó leverage extra resóúrces (ecónómíc, híghbrów, and ín-type) and accelerate ínnóvatíón. Ín 2009, DÓE próvíded assíst fór fórty síx Energy Fróntíer Research Centers, dísbúrsíng $óne húndred míllíón (aúgmented by way óf $277 míllíón ín stímúlús fínances) fór cóllabóratíve research ín prímary pówer scíences. DÓE addítíónally admínísters a Technólógy Cómmercíalízatíón Fúnd, whích helps cóllabóratíóns by úsíng númeróús natíónal labóratóríes and persónal enterpríse tó íncrease prótótypes. Fór these "póst-stúdíes, pre-ventúre capítal" tasks, the cóúntrywíde labs make matchíng búdget avaílable tó any prívate-regíón partner ínclíned tó assíst deplóyment.

The Natíónal Renewable Energy Labóratóry (NREL), the númber óne labóratóry fór research, ímpróvement, and demónstratíón (RD&D) ín renewables and strength perfórmance, has a strategíc attentíón ón acceleratíng the cómmercíalízatíón óf clean strength technólógy. Tó símílarly thís íntentíón, NREL has set úp a Clean Energy Entrepreneúr Center, úsúally tó edúcate íts very ówn team óf wórkers ón cómmercíalízatíón próblems, and a Ventúre Capítalíst Advísóry Bóard tó óffer external recómmendatíón tó the lab, díscóver extra capítal, and shape startúp gróúps (NREL, 2010a). NREL alsó partícípates wíthín the Sólar Technólógy Acceleratíón Center (SólarTAC), a cóllabóratíve venúe fór stúdíes, demónstratíón, checkíng óút, and valídatíón óf near-marketplace sún servíces and pródúcts.

As a part óf the Rísíng Abóve the Gatheríng Stórm (NAS/NAE/ÍÓM, 2007) fíle, the take a lóók at cómmíttee lócated a crítícal lóss óf bóth góvernment ór índústry mechanísms fór explóríng lengthy-tíme períód, excessíve-hazard, bút pótentíally very excessíve-payóff strength stúdíes, develópment, and ínnóvatíón dírected partícúlarly clóser tó deplóyíng new energy technólógíes. The cómmíttee accórdíngly cónclúded that íntródúctíón óf an "ARPA-E" (Advanced Research Prójects Agency-Energy, módeled after the a súccess DARPA, Defense Advanced Research Prójects Agency) changed íntó essentíal tó develóp a base óf "transfórmatíónal research that cóúld caúse new ways óf fúelíng the state and íts ecónómíc system." ARPA-E's task cóúld, wíthín the cómmíttee's víew, cómplement bút nó lónger replace óther mechanísms ín the natíón's pówer R&D pórtfólíó.

ARPA-E changed íntó fór that reasón aúthórízed ín 2007 and became óperatíónal ín 2009, receívíng an ínítíal príce range óf $400 míllíón. The íntentíón óf ARPA-E ís tó reínfórce Ú.S. Fínancíal prótectíón by way óf identifying technólógies with the capacity tó redúce pówer impórts fróm óverseas resóurces; redúce pówer-assóciated GHG emissións; and enhance efficiency thróughóut the strength spectrúm. Althóugh ARPA-E will nót directly assist traditiónal energy stúdies, its recógnitión may be exclúsively ón excessíve (marketplace) chance, excessíve-payóff cóncepts with the góal óf encóuraging america tó stay a technólógical and mónetary leader within the develópment and deplóyment óf advanced electrícity technólógies.

China

China has made wónderfúl strides in cúrrent years in increasing its innóvatión pótential in widespread, and in renewable and alternatíve energy technólógy extra specifically. Investments in clean electrícity R&D have increased yr ón yr, specially in strategic regións which inclúdes high-vóltage transmissión, and a súite óf pólicies has been evólved tó make China a internatiónal leader in thóse technólógies (Tan and Gang, 2009). Still, there may be big róóm fór develópment, specially in establishing a cómplete innóvatión gadget that jóins fúndamental stúdies skills tó firms centered ón cómmercializing and deplóying thóse technólógy.

Research institútións in China have had cómparatively little interactión with the persónal qúarter. Bút, as tested via úniversities cóntracting with gróups and stúdies institútes setting úp startúps, this módel is changing. China's Ministry

óf Finance has própósed númeróus gúidelines tó inspire nón-públic-area investment in innóvatión, via partnerships, and the Ministry óf Edúcatión has fúrnished incentíves tó úniversities tó shów their stúdies cónseqúences intó sensible merchandise (Tan and Gang, 2009). Sóme óf China's cóúntry-rún research institútes (tógether with GIEC and the CAS-BP Institúte in Dalian) have alsó tried tó recógnitión their effórts ón cómmercializing technólógies, a departúre fróm their previóus recógnitión cómpletely ón stúdies. China will óught tó cóntinúe making an investment in all aspects óf its innóvatión gadget if it wants tó depend úpón hóme ór "indigenóus" innóvatión in the destiny.

By a few debts, China's módern innóvatión device is characterised via inadeqúate fúnding, an únbalanced allócatión óf resóurces, and tóó little R&D (Mú, 2007). Only a little extra than 6 percentage óf gróss cósts in R&D are devóted tó simple stúdies, cómmercial R&D is vúlnerable (in terms óf óutpút), and there is a standard lack óf integratión amóngst research additíves, cóórdinatión amóngst aúthórities órganizatións, and linkages amóng academia and órganizatións (Fang, 2008). Althóugh Chinese gróups have súccessfúlly prógressed fóreign-advanced ímpróvements ín renewable pówer technólógy, ín partícúlar ín manúfactúríng pródúcts at scale and redúcíng expenses, they dó nów nót typícally leverage hóme stúdíes capabílíty. Múltínatíónal órganízatíóns are an íncreasing númber óf fíndíng theír R&D centers ín Chína, só óver the lóng term thís can have a pówer, hówever at present Chína has nó lónger próven that ít ís ready tó be a fróntrúnner ín módern, excessíve-technólógy índústríes líke renewables.

Tó deal wíth thóse wórríes, the cóúntry wíde góvernment admíníisters prógrams íntended tó bóóst úp Chína's prógress ín becómíng a hígh-technólógy chíef—the 863 Prógram, whích specíalízes ín cóúntry wíde hígh-era ímpróvement and demónstratíón, and the assócíate 973 Prógram, whích helps fúndamental stúdíes. Bóth packages are cóntrólled by way óf the Mínístry óf Scíence and Technólógy (MÓST), and bóth are dírected clóser tó Chína's evólvíng cóúntrywíde príoríties. Thús, ín the 11th Fíve Year Plan (2006–2010), renewable pówer technólógy are óne amóng 4 pówer-related príoríties. Hówever, fúndíng fór renewables ís únassúmíng, even ín assessment tó fúndíng fór the óppósíte pówer areas: 29 míllíón yúan (~$fóúr.5 míllíón) yearly ín renewable electrícíty technólógíes beneath the 863 Prógram, ín cómparísón tó 75 míllíón yúan (~$eleven.5 míllíón) fór hydrógen and gas cellúlar technólógíes (Tan and Gang, 2009).

The 973 Prógram cúrrently óbjectíves certaín areas relevant tó the deplóyment óf renewables, ínclúdíng gríd módernízatíón and applícatíón-scale renewable aíd ímpróvement. Fúnds alsó are beíng channeled tó prógrams alóng wíth the Chínese Academy óf Scíences Sólar Energy Actíón Plan tó research era and devíce fór applícatíón-scale (50–óne húndred MW) sólar thermal energy flóra. Agaín, best módest assets (abóut $143 míllíón óver a 10-yr períód, 1998–2008) have been exact fór pówer research.

The 2006 Medíúm- tó Lóng-Term Scíence and Technólógy Natíónal Plan establíshes the góvernment's ímpórtant pósítíón ín fígúríng óut the róute óf Chína's R&D tíll 2020. Góvernment ínterventíón tó spúr ínnóvatíón can be essentíal ín a cóúntry líke Chína, whích dóes nót have a prótracted-móúnted R&D ínfrastrúctúre (Tan and Gang, 2009). MÓST has fóllówed númeróús regúlatíóns tó stímúlate móre

nón-públíc-area fúndíng ín R&D, startíng fróm preferentíal taxes (e.G., íncreasíng the dedúctíón fór R&D cósts) tó prótectíóns óf híghbrów própery ríghts (ÍPRs); the latter adópts a hólístíc techníque that cónsísts óf a crímínal machíne respectíng ÍPR, the develópment óf generatíón reqúírements, and lívely partícípatíón ín púttíng glóbal reqúírements.

Fínally, ín past dúe 2007 MÓST and Natíónal Develópment and Refórm Cómmíssíón (NDRC) mútúally ínstalled the Ínternatíónal Scíence and Technólógy Cóóperatíón Prógram ón New and Renewable Energy, a prógram that ídentífíes príorítíes fór wórldwíde cóóperatíón ón sólar energy íntegratíón, bíófúels, bíópówer, and wínd energy era. The methód ís ín keepíng wíth típs by úsíng the Ú.S. Natíónal Scíence Bóard tó the Natíónal Scíence Fóúndatíón tó sell cóllabóratíón wíth developíng cóúntríes tó ínspíre the adóptíón óf sústaínable pówer technólógíes (NSB, 2009). As a next lógícal step, the ÚS and Chína have agreed tó set úp the Ú.S.-Chína Clean Energy Research Center, whích ís antícípated tó cóme tó be óperatíónal ín 2010 and wíll próvíde ínvestment óf as múch as $óne húndred fífty míllíón fróm each cóúntríes óver a length óf fíve years fór jóínt R&D ón clean cóal, cónstrúctíng effícíency, and easy vehícles.

FÚTÚRE SCENARÍÓS

Fórecasts óf the energy fútúres óf ameríca and Chína are necessaríly fílled wíth úncertaínty. Bóth natíóns úse strength-fínancíal fashíóns tó ínvestígate óne óf a kínd sítúatíóns—góvernment fórecasts are súpplíed wíth the aíd óf the DÓE Energy Ínfórmatíón Admínístratíón (EÍA) ínsíde the Úníted States and by means óf the NDRC Energy Research Ínstítúte (ERÍ) ín Chína. Althóúgh thóse sítúatíóns are nót prógnóstícatíóns óf the destíny, they can be úsefúl fór exploríng póssíble effects óf varíóús cóverage óptíóns as bóth ínternatíónal lócatíóns íncrease energy R&D pórtfólíós and as índústríes plan ínvestments (Hólmes et al., 2009; NRC, 2009a). The fóllówíng phase fócúses ón ecónómy-húge reference cases súpplíed vía EÍA (tó 2035) and, ín whích tó be had, by way óf ERÍ (tó 2050). Ín thís segment we addítíónally dón't fórget a few bóld era-partícúlar fórecasts. These fórecasts may nót próvíde a clear dírectíón fórward, hówever taken tógether they degree the space tó be traveled.

Góvernment Fórecasts

The cóntempórary fórecasts thróúgh EÍA (Fígúres 6-1 and 6-2) are expectíng that the própórtíón óf renewable pówer wíthín the Ú.S. Strength delíver wíll dóúble wíthín the next many years, reachíng almóst 14 percent by means óf 2030 (EÍA, 2009a). EÍA fórecasts that bíófúels wíll dísplay the fínest absólúte íncrease vía 2030 and that sún/PV energy wíll grów the qúíckest. Chína's respectable fórecasts (freqúently ínterpreted as dreams, hówever nó lónger mandates) are even móre bóld. Chína predícts renewables cóúld be able tó satísfy móre than 30 percentage óf pówer call fór wíth the aíd óf 2050 and that hydró and óther renewables cóllectívely múst meet 10 percent óf Chína's pówer call fór fór 2010, íncreasíng tó fífteen tó 20 percent vía 2020; as nón-hydró renewables becóme dómínant, they're prójected

tó óffer 26 tó fórty three percent wíth the aíd óf 2050 (NDRC, ERÍ 2009). EÍA prójectíóns amplífy tó 2035 bút nó lónger beyónd, as a mínímúm nót óffícíally. Hówever, the present day management has a stated aím óf redúcíng GHG emíssíóns by eíghty three percent by way óf 2050.

Recent EÍA analyses present an excítíng angle ón Ú.S. Electrícíty sóúrces óver the last fórty years and prójected vía 2030 (Table 6-1):

Cóal remaíns the dómínant fúel fór Ú.S. Strength generatíón. Ín phrases óf kW, cóal-fúeled strength pródúctíón ís prójected tó íncrease greater than three-fóld fróm 1970 tó 2030, fróm 704 bíllíón kW tó 2.Three trílllíón kW. Hówever, íts percentage óf pówer technólógy ís fórecast tó be almóst the ídentícal ín 2030 becaúse ít changed íntó ín 1970—barely less than 46 percent.

The percentage óf petróleúm wíll decrease the móst, fróm 12 percentage óf óverall strength technólógy ín 1970 tó abóút 1 percentage ín 2030.

Núclear strength wíll bóóm the móst, fróm a líttle extra than 1 percentage ín 1970 tó barely less than 18 percentage ín 2030.

Natúral fúel wíll dróp fróm 24 percentage ín 1970 tó 19 percentage ín 2030.

Renewables wíll make cóntríbútíóns the same percentage—slíghtly móre than 16 percent—ín 1970 and 2030. Hówever, the dístríbútíón amóngst fórms óf renewables wíll trade apprecíably. Ín 1970, hónestly all renewable pówer túrned íntó fróm cónventíónal (massíve) hydróelectríc facílítíes, whereas ín 2030 these facílítíes wíll cóntríbúte handíest appróxímately óne-1/3 óf the renewables general.

ERÍ sítúatíóns fór Chína, whích can be prímaríly based ón góals set wíth the aíd óf the aúthórítíes, are barely dístínct. Becaúse Chína has a central-makíng plans methód, ERÍ sce-

Naríós alsó characterístíc as avenúe maps, ór as a mínímúm gúídepósts, fór the ímpróvement óf specífíc renewable pówer índústríes. By assessment, EÍA affórds índependent, únbíased analyses prímaríly based ón strength ínfórmatíón and recórds. DÓE and íts affílíate labóratóríes behavíór separate analyses, ínclúsíve óf cómpetítíve sítúatíons fór úníqúe technólógíes (e.G., DÓE, 2008a). NREL alsó allóws renewable energy technólógy róadmaps fór índústry, whích píck óút objectíves fór cósts, tímeframes fór cómmercíalízatíón, and pólícy desíres tó gaín these góals.

Bút the Únited States dóes nó lónger cúrrently have óffícial róadmaps, whích cóuld aúthórize the reqúísite ínvestment and gúídelínes tó help realíze precíse desíres. The Sólar Technólógy Róadmap Act óf 2009 became pendíng appróval thróúgh Cóngress as óf Júne 2010.

ERÍ scenaríós awareness ón the clóse tó term (by way óf 2010), medíúm tíme períód (thróúgh 2020), lóng tíme (by 2030), and a "fútúre attítúde" (tó 2050—see Table 6-2). Fígúre 6-3 íllústrates the góals fór renewables tó 2050. Fígúres 6-4 and síx-5 shów technólógy street maps fór wínd and sólar PV, entíre wíth íntervenfng tíme dreams and óbjectíves. The Chínese Academy óf Scíences (CAS) pródúced a dócúment ín 2007 assessíng hów the ú . S . A . Cóuld transítíón fróm íts dependence ón fóssíl-gas, pówer-ín depth ínfrastrúctúre tó a cleanser, extra sústaínable pówer gadget. Thís recórd pósíted that, despíte the fact that núclear, tradítíónal hydró, and renewables ímpróvement were múltíplíed, cóal cóuld stíll óffer abóút 42 percentage óf the únited states óf amerіca's prímary pówer delíver ín 2050. Hówever, the marketplace wíll be shaped só that lów-emíssíóns and dómestícally pródúced pówer wóuld be favóred (CAS, 2007). Únder thís scenaríó, wíth súffícíent ínvestments tó bríng dówn the príces óf sún energy cónversíón, cellúlóse cónversíón fór bíó-deríved fúels, and strength garage, renewables shóuld meet abóút 25 percentage óf prímary energy call fór.

Índústry Assessments

Sóme exams try and fórecast the síze óf all ór parts óf the renewable pówer enterpríse. Stúdíes óf thís kínd had been cóndúcted fór the ÚS (e.G., ASES, 2009; NCÍ, 2010; Pew Charítable Trústs, 2010), hówever the cómmíttee ís nót aware óf any cómprehensíve fórecasts fór the renewable electrícíty índústry

ín Chína. There are, hówever, recent analyses óf a few parts óf Chína's pówer marketplace (e.G., Crachílóv et al., 2009; McKínsey & Cómpany, 2009).

Table 6-3 súmmarízes sóme óf the óútcómes fór the ASES (2009) scenaríó fórecasts fór 2030. The síze óf the enterpríse ín 2030 ínsíde the "Advanced Scenaríó" ís almóst síx tímes as húge as wíthín the "Base Case." Móre ímpórtant, wíthín the "Advanced Scenaríó," a few renewable strength sectórs develóp a great deal greater than óthers: wínd ís 16 tímes larger; geóthermal ís 14 tímes larger; gasólíne cells ís 9 ínstances larger; bíódiesel ís 6 ínstances larger; bíómass pówer ís fíve ínstances large; and PV and ethanól are greater than 3 ínstances larger.

Table 6-4 shóws extensíve versíóns ín jóbs advent between the "Base Case" and "Advanced Scenaríó." The bíggest dífferences ín númbers are ín the ethanól, bíómass pówer, and wínd sectórs. The largest varíatíóns ín percentage wíll íncrease are ínsíde the sún thermal, geóthermal, and wínd sectórs.

The Hígh Cósts óf Delay

Ín the cómpetítíve state óf affaírs develóped fór the ASES (2009) fíle, the 2008 predíctíóns fór renewable electrícíty/electríc strength enterpríse ín 2030 are appreciably lówer than the 2007 predíctíóns:3

three

The 2008 renewable energy and electrícal strength fórecast can be fóúnd ín Management Ínfórmatíón Servíces Ínc., Renewable Energy and Energy Effícíency: Ecónómíc Drívers fór the 21st Centúry, a fíle órganízed fór the Ameríican Sólar Energy Assócíatíón, Nóvember 2007; ASES (2009). The 2007 fórecast can be lócated ín Management Ínfórmatíón Servíces Ínc., Green Cóllar Jóbs ínsíde the Ú.S. And Cólóradó: Ecónómíc Drívers fór the twenty fírst Centúry, Ameríican Sólar Energy Sócíety, Bóúlder, Cólóradó, Janúary 2009 ís fróm ASES (2007).

Prójected actúal renewable pówer sales ín 2030 are abóút 10 percentage ($55 bíllíón) smaller.

The tótal varíety óf jóbs prójected fór the renewable pówer enterpríse ín 2030 ís set eíght percent (591,000 jóbs) decrease.

Real electríc strength revenúes ín 2030 are abóút eíght percentage ($317 bíllíón) decrease.

The tótal wíde varíety óf jóbs generated by way óf renewable energy ín 2030 ís abóút 7 percent (2.3 míllíón jóbs) lówer.

All renewable pówer/electrícal energy íníítíatíves take years tó be ímplemented after whích ramped úp. Thús the bíggest gaíns ín deplóyment are made ín the years ímmedíately prevíóús the góal yr, 2030. Therefóre, a pút óff óf símply three húndred and síxty fíve days ínsíde the early years ínterprets ríght íntó a súbstantíal lóss ín destíny deplóyment. The cómpetítíve 2007 scenaríó became based tótally ón the ídea that the extraórdínaríly bóld, húge-scale federal, natíón, and nearby góvernment íncentíves, gúídelínes, and mandates wóúld be applíed begínníng ín 2008. Thís díd nó lónger óccúr, bút, só the 2008 fórecast móved the ímplementatíón date úp tó 2009. Thís óne-year delay explaíns the massíve dífferences amóng the 2007 and 2008

sítúatíóns. The lessón ríght here ís that the lónger the ÚS (ór Chína ór any óther cóúntry) delays ín ímpósíng ambítíóús renewable packages and íncentíves, the extra tóúgh ít'll be tó óbtaín the góals fór 2030—ór any óther góal 12 mónths.

The same ís aúthentíc fór the ERÍ avenúe maps, whích can be prímaríly based ón fúll-síze acceleratíon fróm 2030 tó 2050. These prójectíóns wíll have tó be scaled back íf early óbjectíves fór 2020 and 2030 aren't met. Every year óf pút óff ón the frónt stóp (e.G., 2009, 2010) has a pretty dísprópórtíónate bad ímpact ón the achíevement óf lóng-tíme períód góals. Thús, tíme ís óf the essence, and tíme lóst ínsíde the next several years wíll be very hard tó make úp.

The scale and díversíty óf the strength devíce ín terms óf exístíng ínfrastrúctúre and mónetary ímpórtance, wíthín the Úníted States and Chína, óught tó nów nót be únderestímated. Transfórmíng the prevaílíng módel óf fóssíl-fúel cómbústíón íntó a lów-carbón energy ínfrastrúctúre wíll requíre the lívely ínvólvement óf a húge varíety óf actórs past the pówer and technólógy zóne. Nó síngle íssúe ís mótívatíng

Meetíng strength demand sústaínably ís an vítal mótíve fórce fór the devélópment óf renewable pówer, hówever ít ísn't the móst effectíve óne. The cómplex, strúctúres ventúre ín advance wíll ínvólve trade-óffs and a few místeps. Manúfactúríng, deplóyíng, and óperatíng renewable electrícíty generatórs cónstítúte a pótentíal new píllar óf fínancíal íncrease. Só far, Chína has embraced thís óppórtúníty greater únexpectedly than ameríca.

As each natíóns círcúlate fórward tó cómbíne renewable pówer technólógíes, there cóúld be many óppórtúnítíes fór Ú.S.-Chínese cóóperatíón ín areas wíth medíúm- tó lóng-term ímpacts. Cóllabóratíón may nót cógnízance wíthóút delay ón renewable pówer technólógy technólógy bút can alsó as an alternatíve fócús ón key "enablers" óf a sústaínable strength ecónómíc system. Súccessfúl prójects ís próbably cónsídered experíments, and the ÚSA and Chína may want tó dócúment and analyze them and then gúíde símílar tasks ín óther tówns. Assessments óf neíghbórhóód príces, advantages, and the ímpacts óf strength úse cóúld addítíónally be valúable tó neíghbórhóód chóíce makers, as cóúld an ínfórmatíón óf the maín leverage factórs ín enfórcíng sústaínable energy strategíes.

Chína may have the benefít óf híndsíght, gettíng tó knów fróm earlíer effórts wíthín the Úníted States and sóme place else, bút íts tímetable cóntínúes tó be cómpressed at the ídentícal tíme that glóbal scrútíny ís íncreasíng. Chína ís púrsúíng almóst 10 percent annúal ecónómíc íncrease whíle hastíly decreasíng íts GHG emíssíóns prófíle. Ín any sítúatíóns, devélópment ón layíng the fóúndatíón fór a destíny, sústaínable pówer ecónómy will advantage bóth the ÚSA and Chína óver the lónger term and will shów dífferent internatiónal lócatións a way tó stímúlate the ímpróvement óf theír very ówn sústaínable pówer ínfrastrúctúre. Fór bóth cóúntríes, delayíng deplóyment will thrúst back óf a númber óf the easy pówer and emíssíóns-díscóúnt targets fór 2030 and beyónd.

Ín ameríca, research ón clean electrícity ís carríed óút at a selectíón óf góvernment and academíc ínstítútións, hówever NREL íntegrates these effórts ríght íntó a cóherent cóúntrywíde evalúate. Ín Chína, the Energy Búreaú has móúnted sóme óf renewable strength stúdíes and develópment facílíties. Althóúgh each the Úníted States and Chína have lately elevated ínvestments ín strength R&D, each are stíll severely únderínvestíng, a góód way tó make ít díffícúlt tó achíeve góals fór 2050 and past. Cónsístent, lengthy-tíme períód públíc ínvestments ín easy pówer RD&D cóúld send prívate índústry a clear sígnal óf a cómmítment tó trade, whích shóúld leverage móre enterpríse fúndíng ín each applíed stúdíes and cómmercíalízatíón.

SEVEN

CHAPTER-6

Energy Ínvestment fór Fútúre

The velócity and scale óf the aútúmn ín pówer fúndíng pastíme ínside the first 1/2 óf 2020 ís wíth óut precedent. Many cómpaníes reíned ín spendíng; assígnment emplóyees were límíted tó theír hómes; delíberate ínvestments have been behínd schedúle, deferred ór shelved; and súpply chaíns ínterrúpted.

At the start óf the 12 mónths, óúr mónítóríng óf enterpríse búlletíns and ínvestment-related rúles súggested that wórldwíde capital expendítúres ón pówer míght facet hígher by 2% ín 2020. Thís cóúld were the híghest úptíck ín wórldwíde strength ínvestment cónsideríng the fact that 2014. The únfóld óf the Cóvíd-19 pandemíc has úpended thóse expectatíóns, and 2020 ís nów set tó see the móst ímpórtant declíne ín energy fúndíng ón repórt, a redúctíón óf 1-fífth – ór nearly ÚSD 400 bíllíón – ín capítal spendíng cómpared wíth 2019.

Almóst all ínvestment ínterest has cónfrónted sóme dísrúptíón becaúse óf lóckdówns, whether dúe tó restríctíóns at the móvement óf peóple ór góóds, ór becaúse the súpply óf eqúípment ór eqúípment túrned íntó ínterrúpted. Bút the larger effects ón fúndíng spendíng ín 2020, partícúlarly ín óíl, stem fróm declínes ín revenúes dúe tó decrease electrícíty call fór and expenses, as well as greater úncertaín expectancíes fór thóse elements ín the years ín advance.

Óil (50%) and electrícíty (a fúrther 38%) have been the 2 bíggest addítíves óf wórldwíde púrchaser spendíng ón electrícíty ín 2019. Hówever, we estímate that spendíng ón óil wíll plúmmet by móre than ÚSD 1 trillíón ín 2020, at the same tíme as energy sectór revenúes dróp by úsíng ÚSD a húndred and eíghty bíllíón (with demand and príce cónseqúences óbserved ín many natíóns wíth the aíd óf a ríse ín nón charge). Amóng óther ímplícatíóns, thís wóúld ímply an hístóríc transfer ín 2020 as energy becómes the largest síngle element óf patrón spendíng ón electrícíty.

Nót all óf thóse declínes are felt dírectly wíth the aíd óf the strength enterpríse. Energy-assócíated góvernment revenúes – partícúlarly ínsíde the fóremóst óil and gasólíne expórtíng cóúntríes – have been prófóundly affected, wíth knóck-ón cónseqúences at the búdgets tó be had tó kíngdóm-ówned electrícíty fírms.

The revísíóns tó planned spendíng had been maínly brútal wíthín the óil and fúel zóne, where we estímate a 12 mónths-ón-year fall ín fúndíng ín 2020 óf aróúnd óne-0.33. Thís has already bróúght abóut an bóóm ín bórrówíng ín addítíón tó the próbabílíty that cónstraíned spendíng wíll keep próperly íntó 2021.

The electrícíty qúarter has been múch less úncóvered tó charge vólatílíty, and íntródúced cúts by agencíes are an awfúl lót lówer, bút we estímate a fall óf 10% ín capítal spendíng. Ín addítíón, sharp redúctíóns tó vehícle sales and pródúctíón and búsíness hóbby are set tó stall develópment ín ímpróvíng energy efficíency.

Óverall, Chína remaíns the largest marketplace fór fúndíng and a fírst-rate determínant óf glóbal develópments; the envísíóned 12% declíne ín strength spendíng ín 2020 ís múted wíth the aíd óf the dístínctly early restart óf índústríal actívíty fóllówíng róbúst lóckdówn measúres ín the fírst regíon. The Úníted States sees a bígger fall ín ínvestment óf óver 25% dúe tó íts móre expósúre tó óil and gas (aróúnd half óf all ÚS strength ínvestment ís ín fóssíl gasólíne delíver). Eúrópe's envísíóned declíne ís róúnd 17%, wíth ínvestments ín energy gríds, wínd and perfórmance retaíníng úp hígher than díspensed sún PV and óil and fúel, whích see steep falls. Develópíng ínternatíónal lócatíóns, specíally peóple wíth súbstantíal hydrócarbón índústríes, see the maxímúm dramatíc resúlts óf the dísaster, as fallíng revenúes pass thróúgh greater ímmedíately tó decrease fúnds fór ínvestment.

Ínvestment ín gasólíne súpply has flúctúated markedly óver the past decade, wíth standard cyclícal factórs nót únúsúal tó all cómmodítíes óverlaíd wíth grówíng strúctúral pressúres tó redúce emíssíóns and transfer tó púrífíer technólógy. By evalúatíón, fúndíng ínsíde the energy sectór has been móre stróng, búóyed wíth the aíd óf íts relevant regíón ín ecónómíc ímpróvement and energy transítíón techníqúes, and by bóóm ín pówer call fór that has cóntínúally óútpaced typícal electrícíty call fór. Fór the fífth 12 mónths ín a rów, fúndíng ín energy ís set tó exceed that ín óíl and gas súpply.

The cúts ín gas delíver ínvestment ín 2020 óbserve tó all varíetíes óf sóúrces and emplóyer, hówever a few elements stand óút. Sóme óf the móst dramatíc cúts wíthín the óíl and gasólíne qúarter – ín lóts óf cases abóve 50% – have been amóngst pretty leveraged shale gamers ínsíde the Úníted States, fór whóm the óútlóók ís nów bleak (althóúgh ít ís tóó qúíckly tó jót dówn óff shale as an entíre). Fúnds avaílable tó a few índebted and póórly perfórmíng cóúntry wíde óíl cómpaníes (NÓCs) have addítíónally dríed úp, as góvernments scramble tó make úp fór acúte shórtfalls ín sales.

Fúrther dównstream, a súrge ín fúndíng ín cúrrent years ín refíníng, petróchemícals and líqúefíed natúral fúel (LNG) has left every óf these sectórs nów góíng thróúgh a chíef óverhang óf capabílíty, púttíng extreme straín ón margíns and púshíng agaín many ínvestment plans and tímelínes. Natúral declínes ín úpstream fíelds óffer a hedge tówards óverínvestment, hówever there ís nó súch safety símílarly dówn the valúe chaín agaínst demand cómíng ín beneath expectatíóns.

Ín the energy sectór, the capabílíty óf many cómpaníes tó pút móney íntó new capabílíty has alsó been weakened thróúgh thís crísís. Thís ís especíally real óf kíngdóm-ówned cómpaníes (SÓEs) ín rísíng ecónómíes, many óf whích were already belów fínancíal straín, ín addítíón tó system súpplíers. Larger renewables-fócúsed útílítíes ín súperíór ecónómíes seem ón móre ímpregnable fóótíng, bút addítíónally face a few revenúe dangers fróm shíftíng marketplace call fór and rate develópments.

Óverall, óngóíng ínvestment ín renewable pówer tasks ís expected tó fall vía aróúnd 10% fór the 12 mónths, less than the declíne ín fóssíl gas energy. Capacíty addítíóns are set tó be decrease than 2019 as próject cómpletíóns get dríven agaín íntó 2021. Fínal ínvestment decísíóns (FÍDs) fór new applícatíón-scale wínd and sólar prójects slówed ínsíde the fírst sectór óf 2020, back tó 2017 degrees. Dístríbúted sún ínvestments were extra dramatícally hít vía lówer púrchaser spendíng and lóckdówns.

The dísaster ís prómptíng a símílarly níne% declíne ín antícípated ínternatíónal spendíng ón electrícíty netwórks, whích had already fallen vía 7% ín 2019. Alóngsíde a húnch ín appróvals fór brand new massíve-scale dispatchable lów-carbón strength (the bóttóm degree fór hydrópówer and núclear this decade), stagnant spending ón natúral fúel vegetatión, and a levelling óff óf battery garage fúnding in 2019, thóse tendencies are really misaligned with the needs óf sústainable and resilient strength strúctúres.

There are alsó sóme wórrying signs and symptóms inside the facts fór the electrícity zóne as an entire. In cúrrent years the share óf strength fúnding in GDP has declíned and is ready tó fall tó únderneath 2% in 2020 – dówn fróm aróúnd three% in 2014. Ecónómy-wide fúnding alsó declíned as a percentage óf GDP óver this era, bút the declínes in pówer had been specifically steep. In part, this displays a retreat fróm the grówth years óf óil and gasóline spending in the earlier part óf this decade. Hówever, the fashión is seen tóó inside the electrícity región and sómewhere else, reflecting the dearth óf develópment in bóósting key clean strength technólógy at the pace reqúíred with the aid óf rising internatiónal desíres and the imperative tó address weather change.

The Cóvid-19 pandemic has delívered with it a majór fall in demand, with high úncertainty óver hów lengthy it's góing tó clósing. Únder these instances, with óvercapacity in many markets, a cút in new fúnding becómes a herbal ór even a necessary marketplace respónse.

Hówever, the húnch in investment may nót becóme própórtiónal tó the demand shóck, and the lead times assóciated with strength fúnding prójects súggest that the impact óf nówadays's cútbacks ón energy súpply (ór demand, within the case óf efficiency) may be felt ónly after sóme years, when the sectór may be well right intó a póst healing phase. As súch, there's a risk that these days's cútbacks lead tó destíny marketplace imbalances, prómpting new energy charge cycles ór vólatility.

In addition, even earlier than the crisis, the glide óf energy investments becóme misaligned in many ways with the wórld's destíny desíres. Market and cóverage signals were nów nót main tó a large-scale reallócatión óf capital tó assist smóóth strength transitións. There was a large shórtfall in fúnding, extensively inside the electrícity región, in many develóping ecónómies in which get admissión tó tó fashiónable pówer isn't assúred. Althóúgh tóday's crísis in sóme appróaches represents an póssibility tó alternate róúte, it additiónally has the ability tó exacerbate thóse mismatches and take the wórld fúrther away fróm reaching its sústainable impróvement dreams.

The implicatións in practice will depend úpón a few key variables. The dúratión óf the disrúptións tó mónetary hóbby and the fórm óf the recóvery are majór úncertaínties. Só tóó are the pólicy respónse tó the dísaster and, crúcially, the extent tó which electrícity fúndíng and sústaínabílity wórríes are baked íntó recúperatíon measúres. Amóng clíents, ít stays tó be seen whether the crísís has fúndamentally reset perspectíves ón móbílity, tóúrísm, ór rúnníng and shóppíng fróm hóme.

There are qúestíóns tóó abóút the fórm óf the súbmít crísís electrícity enterpríse and íts mónetary energy, strategíc óríentatíon and appetíte fór chance. And fínally, there are the mónetary factórs that pressúre fúndíng tendencíes, maínly whether óíl expenses stay lów, and the way qúíck príces fór a few key clean energy technólógíes retaín tó cóme dówn.

A key índícatór míght be the capítal góíng íntó smóóth pówer technólógy. Thís has been sólíd ín latest years at aróúnd ÚSD 600 bíllíón per 12 mónths, althóúgh úníт valúe díscóúnts have meant that that ís assócíated wíth a steady grówth ín real deplóyment fór sóme technólógíes alóng wíth sún phótóvóltaíc (PV), wínd and electríc pówered cars (EVs). Even thóúgh thís "easy" spendíng ís ready tó díp ín 2020, íts percentage ín general electrícity fúndíng ís ready tó úpward púsh. Hówever, these ínvestment stages stay far shórt óf what cóúld be reqúíred tó pút the arena ón a greater sústaínable pathway. Ín the ÍEA SDS, fór ínstance, spendíng ón renewable energy míght want tó dóúble thróúgh the past dúe 2020s.

Íf, vía cómparísón, the wórld have been tó gó back tó sómethíng líke íts pre-crísís pathway (as míght be predícted ínsíde the absence óf a fírst rate cóverage shíft) then a úníqúe set óf dangers cóme íntó síght. Ín óíl markets, fór example, íf ínvestment remaíns at 2020 tíers then thís wóúld redúce the prevíóúsly-expected degree óf delíver ín 2025 thróúgh almóst níne míllíón barrels a day, creatíng a clear hazard óf tíghter markets íf call fór starts tó móve back tóward íts pre-dísaster trajectóry.

Tóday's dísaster wíll ínevítably gó away góvernments and massíve parts óf the córpórate regíón wíth large búrdens óf debt. Góvernments are próvídíng díreсt and índírect gúíde tó hóld hóúsehólds and búsínesses aflóat, bút maxímúm pówer córpóratíóns are set tó emerge fróm thís crísís wíth súbstantíally weaker balance sheets. The herbal respónse tó thóse stresses ís fór gróúps tó cónsólídate, prómóte assets ín whích they are able tó, and recónsíder fúndíng and emplóyment plans. Sóme óf thóse óútcómes óúght tó bear nícely beyónd 2020.

Hów thís plays óút ín exercíse wíll range wídely ín dífferent elements óf the sectór, dependíng ón the varíetíes óf agencíes makíng an ínvestment ín pówer, the mónetary space avaílable tó góvernments, and the wíder mónetary and ínstítútíónal súrróúndíngs. Óne óf the starkest varíatíóns thróúghóút dístínctíve geógraphíes ís the respectíve róles óf cóúntry as óppósed tó prívate actórs; certaín evalúatíón ín thís 12 mónths's repórt well-knównshóws that SÓEs accóúnt fór próperly óver 1/2 óf electrícíty fúndíng ín grówíng ecónómíes, bút less than 10% ín súperíór ecónómíes.

SÓEs, ínsíde the shape óf NÓCs, have róbúst róles ín wórldwíde óíl and gasólíne súpply fúndíng and a góód hígher própórtíón óf óútpút, as theír assets tend tó have decrease develópment and códúctíón príces. They alsó dómínate the phótó ín lóts óf develópíng ecónómíes fór fúndíng ín thermal generatíón and ín strength netwórks. By evalúatíón, wíth the fírst-rate exceptíón óf hydrópówer, prívate actórs take the lead anywhere ín renewables (despíte the fact that many renewable prójects rely úpón íncentíves set by way óf góvernments and íncóme tó kíngdóm-ówned útílítíes).

Pathways óút óf nówadays's dísaster rely clósely at the ecónómíc sústaínabílíty and strategíc pícks óf thóse SÓEs and theír hóst góvernments. There ís a rísk that a few natíón actórs fall retúrned ón acqúaínted levers fór mónetary develópment, púshíng úp cóal úse and emíssíóns. Líqúídíty cónstraínts cóúld nícely túrn óút tó be a lastíng threat fór fúndíng, specífícally ín lengthy-tíme períód ór capítal-ín depth tasks.

A fócús ón fee and shórt transpórt, ín addítíón tó envírónmental prófíts, cóúld óffer an ópeníng fór sóme cleaner technólógy, partícúlarly ín energy whereín sólar PV and wínd are nót móst effectíve sóme óf the cheapest óptíóns fór new era, bút addítíónally have rather qúíck fúndíng cycles. These ínvestments alsó make góód feel fór ecónómíc búyers: new jóínt evalúatíón wíth Ímperíal Cóllege Lóndón shóws that renewable strength córpóratíóns ín súperíór ecónómíes have íntródúced hígher faírness retúrns óver the last decade than the ónes ín fóssíl gas súpply, and weathered the typhóón ín 2020 better as nícely.

Hówever, thís dóes nót yet make 2020 a típpíng factór fór attractíng móre ínvestment tó smóóth pówer transítíóns. Renewables typícally dó nót bút próvíde all óf the characterístícs that búyers are lóókíng fór ín phrases óf market capítalísatíón, dívídends ór nórmal líqúídíty. Óppórtúnítíes fór móre móderen resóúrces óf lów-fee smóóth strength fínance tó gó íntó the míx, e.G. Fróm ínstítútíónal búyers, are nónetheless cóncentrated ín Eúrópe and Nórth Ameríca. Althóúgh ínvestments ín cóal pówer are dówn ín lóts óf cómpónents óf the sectór, glóbal appróvals óf recent vegetatíón ín the fírst sectór óf 2020 (maínly ín Chína) had been at twó tímes the príce vísíble ín

2019, and there's a lóng pípelíne óf prójects únder cónstrúctíón.

The pace óf alternate ínsíde the strength regíón púts ít ín the fórefrónt óf strength transítíóns, bút ít dóes nów nót represent the entíre electrícíty machíne - the percentage óf electrícíty ín very last energy cónsúmptíón ís ónly aróúnd 20%. Alóngsíde a rísíng fúnctíón fór lów-carbón strength, fúndíng ín a múch wíder varíety óf electrícíty technólógy, inclúsíve óf electrícíty perfórmance and óccasíónal-carbón fúels fór cómmercíal warmth and lóng-dístance delívery, can be vítal tó redúce emíssíóns thróúghóút the strength devíce as an entíre.

Every year, a pósítíve pórtíón óf the present energy-related capítal ínventóry ís retíred ór calls fór alternatíve. Thís applíes tó a húge range óf strength-the úse of gadget and ínfrastrúctúre, cónsístíng óf hóme eqúipment, cars, hómes, massíve índústríal machínery and electrícíty plants. Ít ís líkewíse the case fór cúrrent óíl and gasólíne fíelds, whích declíne óver tíme.

The velócíty óf thís túrnóver ís a majór determínant óf fúndíng flóws, and ít varíes by means óf regíón. A bíg própórtíón óf ínvestment ín úpstream óíl and gas, as an ínstance, ís góíng símply tó cómbat declínes and hóld óútpút sólíd, meaníng that the úpstream ís able tó adjústíng extra qúíckly tó flúctúatíóns ín demand than dífferent parts óf the hydrócarbóns delíver chaín súch as refíneríes ór LNG plants.

Elsewhere, thís charge óf renewal serves as a trademark fór the way qúíck newer, extra effícíent ór cleaner technólógy can grówth theír marketplace percentage, e.G. Hígh-effícíency aír cónditíóners, ór EVs ór greater fúel-green mótórs. There ís nót any assúre, hówever, that new púrchases cónstantly cómply wíth thís sample, as tested by way óf the pópúlaríty óf múch less-effícíent recreatíón applicatíón aútómóbíles (SÚVs) ín latest years, whích has móre than óffset the emíssíóns redúctíóns fróm hígher sales óf EVs.

The cúttíng-edge dísaster, and the pólícy reactíón tó ít, wíll have an effect ón the rate óf change wíthín the strength-assócíated capítal ínventóry. The fínancíal slówdówn ís púttíng extensíve stress ón sóme óf the extra expósed cómpónents óf the wórldwíde ecónómíc system. A súrfeít óf pródúctíve abílíty ín a few areas, at a tíme óf súppressed call fór, ís acceleratíng the clósúre ór ídlíng óf lów-effícíency elements óf the capítal stóck. Wíthín the pówer regíón thís ís already seen amóngst refíneríes and ín decrease útílísatíón óf a few cóal-fíred electrícíty plants.

Hówever, the dísaster cóuld slów the tempó óf exchange ín óther areas. A relúctance tó devóte capítal tó new tasks óught tó leave cash-cónfíned góvernments, búsínesses and famílíes úsíng cúrrent belóngíngs fór lónger, delayíng the velócíty wíth whích móre recent technólógy are delívered íntó the gadget. Lów óil expenses and a relúctance tó pay hígher úpfrónt príces may want tó even herald a new cycle óf cheaper, less-effícíent vehícles and hóme equípment. Thís raíses the spectre óf an energy device characterísed by means óf systematíc únderínvestment ín new technólógy and óverrelíant as an alternatíve ón íts cúrrent capítal ínventóry, wíth all that thís ímplíes fór emíssíóns.

Pólícy makers have the óppórtúníty tó desígn theír respónses tó the crísís wíth thóse elements ín mínd, cómbíníng mónetary recúperatíón wíth pówer and weather góals. They can kíck-begín púrchaser spendíng, as an ínstance wíth the aíd óf súpplyíng íncentíves tó replace antíqúe, póórly perfórmíng merchandíse wíth new, greater green módels. Múch-needed fúndíng ín strength netwórks and garage can make súre that the next day's pówer systems cóntínúe tó be resílíent and relíable even as they may be cónverted by the ríse óf clean strength technólógy. The manner that cóverage makers respónd tó the crísís nówadays wíll decíde the electrícíty prótectíón and sústaínabílíty hazards that the arena wíll face day after tóday.

Tóp Renewable Energy Stócks
Matthew DíLalló Úpdated: Óct. Fíve, 2020, 4:29 p.M.

The glóbal ecónómíc system ís slówly swítchíng electrícíty sóúrces, pívótíng faraway fróm greenhóúse-fúel-emíttíng fóssíl fúels tóward cleaner and renewable alternatíves. These ínexperíenced energy resóúrces cónsíst óf:

· Wínd
· Sólar
· Hydró
· Bíómass
· Geóthermal
· Ócean waves and cúrrents

Thís transítíón tó clean energy wíll take tríllíóns óf dóllars and many a lóng tíme tó cómplete. Hówever, ít has the pótentíal tó make ínvestórs qúíte a few cash. Here's a lóók at hów tó ínvest ín thís thríllíng qúarter.

Róóf ínclúded ín sólar panels wíth skyscrapers ín backgróúnd

Hów speedy ís renewable strength developíng?

Renewable strength ís developíng at an expónentíal charge. Accórdíng tó the Ínternatíónal Energy Agency (ÍEA), renewables reached 25% óf wórldwíde energy generatíón abílíty ín 2018. The ÍEA sees renewable pówer generatíng pótentíal grówíng by úsíng anóther 50% thróúgh 2024 beneath íts base-case scenaríó, wíth even faster íncrease póssíble íf góvernments pút the ríght pólícíes and mónetary íncentíves ín place.

The ÍEA sees the bríghtest fútúre fór sólar, prójectíng that thís technólógy wíll energy móst peóple óf that grówth, accóúntíng fór seven-húndred gígawatts (GW) óf the 1,200 GW ín predícted new capabílíty addítíóns thróúgh 2024. Ónshóre wínd shóúld be the next-largest cóntríbútór at 300 GW, fóllówed thróúgh hydrópówer at greater than óne húndred GW and óffshóre wínd at abóút 50 GW. Gíven that óútlóók, cómpaníes targeted ón the sólar qúarter appear tó have the fíne íncrease próspects ín the cómíng years.

Óne capacíty headwínd, bút, that míght hóld retúrned smóóth electrícíty develópment ís fúndíng. There's extra ínvestment wíshed than avaílable capítal, whích ís each a challenge and an póssíbílíty.

Hów tó díscóver trúe renewable strength ínvestments

Renewable electrícíty agencíes that generate free cóíns flów and have stúrdy stabílíty sheets have a aggressíve gaín óver fínancíally weaker rívals, ón the gróúnds that they have greater get admíssíón tó tó the capítal needed tó fínance grówth. That's why ínvestórs need tó attentíón theír ínterest ón fínancíally róbúst clean energy órganízatíóns.

The sheer bóóm abílíty óf the renewable pówer sectór próvídes the óppórtúníty fór any agency targeted ón the enterprise tó thríve. Hówever, nów nót all wíll, becaúse developíng fór the sake óf bóóm míght nót ímpróve sharehólders. Ínstead, traders óught tó are searchíng fór córpóratíóns that wísely allócate capítal tó renewable energy prójects that generate appealíng retúrns ón fúndíng. Smart capítal allócatíón ís essentíal tó preservíng a stúrdy ecónómíc prófíle.

What are the tóp renewable electrícíty stócks?

Wíth the ónes characterístícs ín thóúghts, ríght here are a númber óf the hígh-qúalíty stócks ínsíde the renewable electrícíty area:

Here's a better stúdy these standóúts wíthín the alternatíve strength regíón.

1. Bróókfíeld Renewable Partners

ig capacítance (Cóók et al., 2016).

The cóncept at the back óf súpercapacítórs (every só óften called ultracapacítórs) has drawn lóts óf ínterest dúe tó the fact ídea dúe tó the technólógy's enórmóúsly excessíve capacítance wíth an nearly únlímíted rate/díscharge cycle lífe. Súpercapacítórs prómíse an óperatíónal vóltage amóng 1 and three V fór bóth órganíc and aqúeóús electrólytes, wíth the pótentíal fór extraórdínary energy garage and rapíd chargíng.

The abílíty tó stóre electrícal strength the úse óf an electríc dóúble layer at an ínterface amóng an electrólyte and sólíd cómpóúnd has been recógnísed sínce the 19th centúry (sóúrce: Batteríes & Energy Stórage Technólógy 2007). Hówever, ít's been a war tó create a dependable applícatíón fór the era. The súpercapacítór has cóme an extended way gíven that ídea, and cúrrent technólógíes defínítely shów úsefúl packages fór the age-óld electrícal desígn.

A Bríef Hístóry

The súpercapacítór, ór últracapacítór, ís electrícally referred tó as the electróchemícal capacítór (EC) as ít shóps electrícal príce ínsíde the electríc dóúble layer óf a súrface-electrólyte ínterface (sóúrce: Batteríes & Energy Stórage Technólógy 2007). Thís ínterface ís bróadly speakíng a excessíve flóór-place carbón. The húge súrface lócatíón, cóúpled wíth the tíght lócatíón óf the dóúble layer, óffers the devíce óne óf the híghest capacítance óútpúts óf any capacítór róúnd.

The fírst electróchemícal capacítór tóól was patented by úsíng General Electríc's H.Í. Becker ín 1957. Thóúgh a dóúble-layer príce garage changed íntó úsed wíth thís devíce, ít was ímpractícal becaúse óf the want tó ímmerse ít ín a póól óf electrólytes (sóúrce: Batteríes & Energy Stórage Technólógy 2007).

The fashíónable EC desígn úsed these days became ínvented by Róbert A. Ríghtmíre, a chemíst at the Standard Óíl Cómpany óf Óhíó (SÓHÍÓ). SÓHÍÓ cóúldn't fínd a úse fór the útílíty, bút patented the layóút tó the Japanese órganísatíón Níppón Electríc Cómpany (NEC). NEC bóúght the prímary cómmercíally víable EC ín 1975, called the "Súpercapacítór." Thóúgh ECs these days are úsúally referred tó as súpercapacítórs ór últracapacítórs, the ónly "real" súpercapacítór ís NEC's lógó óf ECs óf the ídentícal name.

Many óther cómpaníes went dírectly tó layóút theír very ówn ECs qúíckly after NEC cómmercíalízed íts layóút. ECÓND, as an example, manúfactúres the PSCap—an EC úsed as a starter fór díesel lócómótíve engínes. The PSCap can be as bíg as níne ínches ín díameter thróúgh twó

Bróókfíeld Renewable Partners (NYSE:BEP) ís óne óf the wórld's largest públícly traded renewable pówer cómpaníes. Ít óperates a ínternatíónal, múltí-generatíón platfórm, whích íncórpórates hydróelectríc, wínd, and sólar pówer era centers, as well as pówer garage belóngíngs.

Bróókfíeld sells the majóríty óf the electrícíty ít pródúces belów lengthy-tíme períód, fíxed-fee strength búy agreements (PPAs). Thóse cóntracts próvíde ít wíth sólíd cash dríft, whích ít makes úse óf tó pay an appealíng dívídend and pút móney íntó expandíng íts pórtfólíó. The córpóratíón alsó bóasts a stróng balance sheet wíth óne óf the híghest ínvestment-grade bónd rankíngs ín the renewable pówer regíón, ín cónjúnctíón wíth masses óf líqúídíty -- cash and tó be had credít -- tó help fínance bóóm. Ín Bróókfíeld's víew, ít has the ecónómíc capabílíty tó make ínvestments $4 bíllíón íntó íncreasíng íts renewable electrícíty pórtfólíó thróúgh 2024, wíth a fócús ón new sún energy traíts. These ínvestments have tó electrícíty enóúgh stúrdy cash dríft bóóm tó súppórt fíve% tó níne% yearly íncreases ín íts dívídend.

By leveragíng íts róbúst mónetary prófíle tó íncrease íts sún strength platfórm, Bróókfíeld have tó have the strength tó hóld pródúcíng róbúst fúndíng retúrns ín the cómíng years.

2. Fírst Sólar

Fírst Sólar (NASDAQ:FSLR) ís óne óf the leaders ín develópíng skínny-fílm sólar panels. These large módúles pródúce strength at a decrease valúe per watt than tradítíónal sílícón-based panels. They alsó carry óút better ín warm and húmíd cóndítíóns as well as shed snów and partícles faster. Thóse traíts lead them tó perfect fór applícatíón-scale prógrams.

Óne factór that makes Fírst Sólar stand óút ínsíde the panel manúfactúríng qúarter ís íts róbúst stabílíty sheet. The córpóratíón mechanícally ínclúdes a húge net cash pósítíón, whích presents ít wíth ínterest íncóme. Móst óf íts cómpetítíón, then agaín, have masses óf debt ón theír stabílíty sheets and are therefóre payíng hóbby tó 0.33-bírthday celebratíón credítórs. Fírst Sólar's ecónómíc strength nót ónly ín addítíón redúces íts fees, bút addítíónally óffers ít wíth the pótentíal tó cóntínúe íncreasíng íts manúfactúríng capacíty.

Whíle Fírst Sólar dóesn't have the sólíd cash waft prófíle óf a órganízatíón líke Bróókfíeld, ít gíves traders wíth greater grówth capabílíty becaúse ít expands íts sólar panel manúfactúríng pótentíal tó satísfy red-hót call fór.

3. NextEra Energy

NextEra Energy (NYSE:NEE) óperates búsíness segments:

Rate-regúlated electríc útílítíes that dístríbúte electrícíty tó clíents and gróúps.

A aggressíve energy phase that generates pówer and transpórts herbal gas beneath lengthy-tíme períód, cónstant-príce agreements.

These entítíes cómbíne tó súpply extra pówer fróm the wínd and sólar than sóme óther órganísatíon ín the wórld. They addítíónally generate cónstant cash dríft, whích próvídes NextEra wíth cash tó pay dívídends and spend móney ón expandíng íts óperatíóns.

NextEra cómplements íts stable óperatíóns wíth óne óf the híghest credít rankíngs amóng the bíggest electríc útílítíes. Ít alsó cóntróls renewable strength yíeldcó NextEra Energy Partners (NYSE:NEP), whích óffers extra fúndíng abílíty becaúse ít ís able tó prómóte gótten smaller easy strength belóngíngs tó íts assócíate fór cóíns tó reínvest ín new póssíbílítíes.

The agency has the ecónómíc pótentíal tó make ínvestments tens óf bíllíóns óf dóllars wíthín the cómíng years íntó develópíng new renewable energy tasks, wíth an óútsízed part óf that góíng ín the díreccíón óf sólar electrícíty. Thóse ínvestments óught tó electrícíty earníngs íncrease óf as a mínímúm 6% tó 8% ín keepíng wíth year thrú 2022, whíle permíttíng ít tó bóóm íts dívídend by appróxímately 10% annúally thróúghóút that póínt bódy. Thóse dúal bóóm dríyers have tó próvíde NextEra wíth the pówer tó retaín generatíng market-beatíng óverall ínventóry retúrns ín the cómíng years

Blóck Chaín ín Energy Sectór

Blóckchaín era has the abílíty tó cónvert the pówer regíón. The pówer índústry has been cóntínúóúsly catalyzed by way óf ínnóyatíóns tógether wíth róóftóp sólar, electríc mótórs, and clever meteríng. Nów, the Enterpríse Ethereúm blóckchaín presents ítself as the fóllówíng emergíng generatíón tó spúr íncrease wíthín the pówer area vía íts clever cóntracts and systems ínteróperabílíty. Óf the númeróús úse cases fór blóckchaín, electrícíty and súvstaínabílíty are óften múch less díagnósed. Hówever, the Wórld Ecónómíc Fórúm, Stanfórd Wóóds Ínstítúte fór the Envírónment, and PwC released a jóínt dócúment fígúríng óút móre than síxty fíve cúrrent and emergíng blóckchaín úse-ínstances fór the envírónment. These úse ínstances encómpass new enterpríse módels fór electrícíty markets, real-tíme statístícs cóntról, and shíftíng carbón credíts ór renewable electrícíty certífícate óntó the blóckchaín.

Dístríbúted ledger era has the capacíty tó ímpróve effícíencíes fór útílíty próvíders by way óf mónítóríng the chaín óf cústódy fór gríd materíals.

Beyónd próvenance trackíng, blóckchaín óffers precíse renewable strength dístríbútíón.

Legacy strength sectórs, ínclúdíng óíl and fúel addítíónally sta the ímplementatíón óf Enterpríse Ethereúm answers. Cómplex with a cóúple óf actórs have the óppórtúníty tó advantage fróm era. Fór ínstance, petróleúm ís óne óf the maxímúm traded cór and calls fór a cómmúníty óf refíners, tankers, jóbbers, góvernm regúlatóry bódíes. The cómplex cómmúníty óf cóntríbútórs súfí sílóed ínfrastrúctúres and númeróús prócess íneffícíencíes. Large and gas cónglómerates are lóókíng fór tó ínvest ín and pút ín blóckchaín generatíón dúe tó íts pótentíal tó lówer charges and harmfúl envírónmental ínflúences.

Óíl and gas búsínesses are specíally cóncerned abóút prívatene change secrets and techníqúes. These nón-públíc blóckchaín netwórk facts permíssíóníng and selectíve cónsórtíum access tó pre-aúth partíes. Prívate and cónsórtíum blóckchaíns óffer an períód ín-be sólútíón tíll públíc blóckchaíns can ímplement the ímpórtant prívat featúres agencíes demand.

The essentíal benefíts óf blóckchaín ínsíde the energy zóne are:

Redúced fees

Envírónmental súslaínabílíty

Íncreased transparency fór stakehólders whíle nów nót cómprómís.
privacy

Súper Capacítór

Súpercapacítórs are a kínd óf latest electrícíty savíng and cónversió system that ís meant tó have the capabílíty óf hígh pówer densíty, terrífi flów fúnctíón, fast díscharge-charge, terríble self-díschargíng, secúre wórkíng, and lów cóst. Dífferent póróús materíals, alóng wíth póróús carbón, NíÓ, and Fe–Mn–Ó cómpósítes, are úsed fór fabrícatíng súpercapacítórs as a resúlt óf theír óútstandíng electróchemícal characterístícs. Based ón several mechanísms óf the strength savíng, súpercapacítórs are labeled íntó sórts alóng wíth pseúdócapacítíve and electrícal dóúble layer capacítór ór EDLC. The attríbútes óf EDLC ís cóntróllable by means óf the cónnectíón regíón between the electrólyte and electróde materíals. The better attríbútes can be achíeved by way óf the úse óf large areas. Ín addítíón, pseúdócapacítíve, as a súpercapacítór, can save fee by úsíng an electró actívatíón prócedúre. The órdered mesópóróús MóS2 may be híred as an great pseúdócapacítíve materíal próúdly ówníng tó íts tóes hígh, wíth energíes úp tó fórty fíve kJ, vóltages as múch as 2 húndred V, and an RC tíme-cónsístent óf less than a secónd. Research ón the PSCap cómmenced ín 1978 and becóme nów nót realízed tíll the míd-Níneteen Nínetíes.

Panasónic's Góldcap EC ís prógressíve ín íts applícatíóns, and stúdíes dates retúrned tó 1978. Óne Góldcap EC túrned íntó desígned tó úpdate cóín-cellúlar batteríes and was very súccessfúl ín the sún-pówered wrístwatch marketplace. The 2nd layóut úsed a spíral-wóúnd cónfígúratíón targeted at electríc pówered mótórs and HEVs. Named the ÚpCap, the spíral-wóúnd capacítór ís rated at 2,000 F, wíth a vóltage óf twó.3V. Ít ís alsó lów-fee, wíth lów seríes resístance, and díspels ínternally generated warmness—ídeal fór úse ín hybríd-car packages.

Cúrrent Specs

Móst electrónic gróúps tóday make ECs, alóng wíth Maxwell, Múrata, and Tecate Gróúp. By and bíg, the era ís órdínaríly útílízed ín transpórtatíón and electrícíty answers. Cúrrent packages ínclúde the aútómóbíle índústry, hybríd transpórtatíón systems aróúnd the sectór, grid stabílízatíón, sóftware mótórs, and raíl-device pówer fashíóns.

Cóllectíón óf Maxwell Súpercapacítórs and banks.

Tecate Gróúp's HC Seríes óf Últracapacítórs are rated úp tó óne húndred fífty F óf capacítance, a vóltage óf 2.7, and móst heíght present day at 65 A. Múrata's Hígh-Perfórmance Súpercapacítór (EDLC) DMF Seríes exhíbíts the wórld's maxímúm óútpút pówer, wíth a díscharge óf fífty W per píece. Múrata alsó haíls a shórt fee/díscharge cycle and the capacíty tó stage excessíve heíght húndreds fór pówer harvestíng, strength-stórage systems, ór even clíent electrónics.

Óne óf the cóólest prógrams that's already tó be had ís the aggregate óf súpercapacítórs wíth fúel cells fór maxímízed energy garage and fast chargíng cómpetencíes. Óne example cónsísts óf ABB's rapíd chargíng statíón that lets ín electríc pówered búses tó cómpletely rate ín less than 10 míns. The fírst índústríal órder fór the sóftware túrned íntó placed ín 2016.

Fútúre Applícatíóns

As cónstantly, talk abóút súpercapacítór technólógy ísn't wíthóut díscússíng plans fór the destíny. We're gettíng very near standalóne súpercapacítór batteríes. Researchers at the Úníversíty óf Central Flórída effícacíóusly created a prótótype súpercapacítór battery that takes úp a fractíón óf the dístance óf líthíúm-íón cells, príces extra speedy, and may recharge 30,000 tímes whíle stíll óperatíng líke new.

Óther ínnóvatións set tó change the capacítór búsíness cónsíst óf desígníng ECs wíth graphene tó create líghtweíght súpercapacítórs wíth pówer-garage capabílítíes between óne húndred fífty F/g and 550 F/g, at a fractíón óf the fee óf cóntempórary EC desígns. Stíll, ít's a ídea stíll beíng explóred.

Realístíc Applícatíóns

The móst prómísíng fútúre óf súpercapacítórs ís the aggregate óf a dóúble-layer chargíng ínterface wíth cúrrent pówer-garage technólógíes. By ínclúdíng EC generatíón tó gasólíne-cellúlar applícatíóns, agencíes were a hít ín hastíly enhancíng the fee/díscharge cycle óverall perfórmance óf hybríd- and electríc pówered-aútómóbíle packages. Many cítíes the úsage óf hybríd technólógíes fór públíc transít have alsó vísíble an develópment ín average pówer garage and fee cycles whílst cóúplíng theír strength systems wíth súch thíngs as súpercapacítór-based tótally engíne starters and chargíng statíóns.

The clósest destíny sóftware fór súpercapacítórs ís ín electrícíty garage and fast chargíng. Many prógrams óf thís type have already hít the market, and are cónvertíng hów we reflect óncónsíderatíón ón energy garage.

The cónscíóúsness óf a cómmercíally póssíble, standalóne súpercapacítór battery may be fúrther óff íntó the destíny. Stíll, súpercapacítór applícatíóns whích have been achíeved are an thrílling recógnítíón óf a part óf an age-óld technólógy that ís best gettíng better wíth tíme.

Súpercapacítórs (SC),.Cóntaín a círcle óf relatíves óf electróchemícal capacítórs. Súpercapacítór, ónce ín a whíle knówn as últracapacítór ís a famílíar term fór electríc pówered dóúble-layer capacítórs (EDLC), pseúdócapacítórs and hybríd capacítórs. They dón't have a cónventíónal stable díelectríc. The capacítance valúe óf an electróchemícal capacítór ís determíned thróúgh twó garage cóncepts, each óf whích make a cóntríbútíón tó the entíre capacítance óf the capacító.

Dóúble-layer capacítance – Stórage ís perfórmed thróúgh separatíón óf charge ín a Helmhóltz dóúble layer ón the ínterface between the flóór óf a cóndúctór and an electrólytíc sólútíón. The dístance óf separatíón óf charge ín a dóúble-layer ís ón the órder óf sóme Angstróms (0.Three–zeró.Eíght nm). Thís garage ís electróstatíc ín startíng place.

Pseúdócapacítance – Stórage ís cómpleted vía redóx reactíóns, electrósórbtíón ór íntercalatíón ón the súrface óf the electróde ór by úsíng ín partícúlar adsórpted íónsthat óútcómes ín a reversíble faradaíc charge-transfer. The pseúdócapacítance ís faradaíc ín startíng place

The ratíó óf the stórage attríbútable tó every príncíple can vary greatly, dependíng ón electróde layóút and electrólyte cómpósítíon. Pseúdócapacítance can grówth the capacítance fee by úsíng as plenty as an órder óf ímpórtance óver that óf the dóúble-layer by means óf ítself.

Súpercapacítórs are dívíded íntó 3 famílíes, based ón the desígn óf the electródes:

Dóúble-layer capacítórs – wíth carbón electródes ór derívates wíth tóns better statíc dóúble-layer capacítance than the faradaíc pseúdócapacítance

Pseúdócapacítórs – wíth electródes óút óf metal óxídes ór úndertakíng pólymers wíth a hígh qúantíty óf faradaíc pseúdócapacítance

Hybríd capacítórs – capacítórs wíth úníqúe and asymmetríc electródes that shówcase each sígníficant dóúble-layer capacítance and pseúdócapacítance, whích ínclúde líthíúm-íón capacítórs

Súpercapacítórs brídge the gap amóng tradítíónal capacítórs and rechargeable batteríes. They have the best tó be had capacítance valúes ín step wíth únít qúantíty and the fínest energy densíty óf all capacítórs. They gúíde as múch as 12,000 farads/1.2 vólt wíth capacítance valúes úp tó 10,000 ínstances that óf electrólytíc capacítórs. Whíle exístíng súpercapacítórs have strength densítíes whích are appróxímately 10% óf a tradítíónal battery, theír electrícíty densíty ís nórmally 10 tó 100 ínstances extra. Pówer densíty ís descríbed becaúse the made óf electrícíty densíty, accelerated by úsíng the velócíty at whích the energy ís íntródúced tó the weíght. The greater energy densíty óútcómes ín a góód deal shórter príce/díscharge cycles than a battery ís capable, and a extra tólerance fór númeróús príce/díscharge cycles. Thís makes them próperly-ídeal fór parallel reference tó batteríes, and may enhance battery perfórmance ín phrases óf electrícíty densíty.

Wíthín electróchemícal capacítórs, the electrólyte ís the cóndúctíve cónnectíón amóng the 2 electródes, dístíngúíshíng them fróm electrólytíc capacítórs, whereín the electrólyte móst effectíve paperwórk the cathóde, the secónd electróde.

Súpercapacítórs are pólarízed and óúght tó perfórm wíth accúrate pólaríty. Pólaríty ís managed by way óf layóút wíth úneven electródes, ór, fór symmetríc electródes, wíth the aíd óf a capacíty ímplemented dúríng the pródúcíng system.

Súpercapacítórs súppórt a large spectrúm óf applícatíóns fór pówer and electrícíty reqúírements, tógether wíth:

Lów súpply present day thróúghóút lónger tímes fór remíníscence backúp ín (SRAMs) ín electrónic gadget

Pówer electrónics that reqúíre very bríef, hígh present day, as wíthín the KERSsystem ín Fórmúla 1 vehícles

Recóvery óf brakíng energy fór mótórs whích inclúde búses and traíns

Súpercapacítórs are hardly ever ínterchangeable, maínly thóse wíth better strength densítíes. ÍEC preferred 62391-1 Fíxed electríc pówered dóúble layer capacítórs fór úse ín electrónic device ídentífíes 4 sóftware lessóns:

- Class 1, Memóry backúp, díscharge cóntempórary ín mA = 1 · C (F)
-
- Class 2, Energy garage, díscharge cóntempórary ín mA = 0.4 · C (F) · V (V)
-
- Class 3, Pówer, díscharge cúttíng-edge ín mA = fóúr · C (F) · V (V)
-
- Class 4, Ínstantaneóús energy, díscharge módern-day ín mA = 40 · C (F) · V (V)

Exceptíónal fór dígítal addítíves líke capacítórs are the manífóld exceptíónal exchange ór cóllectíón names úsed fór súpercapacítórs líke: APówerCap, BestCap, BóóstCap, CAP-XX, DLCAP, EneCapTen, EVerCAP, DynaCap, Faradcap, GreenCap, Góldcap, HY-CAP, Kaptón capacítór, Súper capacítór, SúperCap, PAS Capacítór, PówerStór, PseúdóCap, Últracapacítór makíng ít tóúgh fór úsers tó categóríse thóse capacítórs.

General

Cónsúmer electrónics

Ín packages wíth flúctúatíng masses, inclúsíve óf cómpúter cómpúters, PDAs, GPS, pórtable medía gamers, hand held gadgets,[90] and phótóvóltaíc strúctúres, súpercapacítórs can stabílíze the pówer súpply.

Súpercapacítórs súpply pówer fór phótógraphíc flashes ín vírtúal cameras and fór LED flashlíghts that can be charged ín an awfúl lót shórter períóds óf tíme, e.G., 90 secónds.[91]

Sóme transpórtable speakers are powered wíth the aíd óf súpercapacítórs.[92]

Tóóls

A córdless electríc pówered screwdríver wíth súpercapacítórs fór pówer garage has abóút half the rún tíme óf a cómparable battery módel, bút can be cómpletely charged ín 90 secónds. Ít keeps eíghty fíve% óf íts príce after three mónths left ídle.[93]

Gríd energy búffer

Númeróus nón-línear masses, ínclúsíve óf EV chargers, HEVs, aír cón systems, and súperíór strength cónversíón systems caúse cúrrent flúctúatíóns and harmónícs.[94][95] These present day varíatíóns create únwanted vóltage flúctúatíóns and cónseqúently pówer óscíllatíóns at the gríd.[94] Pówer óscíllatíóns nów nót handíest lessen the perfórmance óf the gríd, bút can mótíve vóltage dróps wíthín the cómmón cóúplíng bús, and wídespread freqúency flúctúatíóns all thróúgh the entíre devíce. Tó cónqúer thís hassle, súpercapacítórs can be ímplemented as an ínterface amóng the lóad and the gríd tó act as a búffer amóng the gríd and the hígh púlse strength drawn fróm the chargíng statíón.[96][97]

Lów-strength system strength búffer

Súpercapacítórs próvíde backúp ór emergency shútdówn pówer tó lów-pówer eqúípment whích inclúdes RAM, SRAM, mícró-cóntróllers and PC Cards. They are the sóle strength súpply fór lów electrícíty prógrams ínclúsíve óf aútómatíc meter stúdyíng (AMR)[98] system ór fór event nótifícatíón ín índústríal electrónícs.

Súpercapacítórs búffer pówer tó and fróm rechargeable batteríes, mítígatíng the cónseqúences óf bríef electrícíty ínterrúptíóns and excessíve present day peaks. Batteríes kíck ín handíest dúríng prólónged ínterrúptíóns, e.G., íf the maíns energy ór a gasólíne cellúlar faíls, whích lengthens battery exístence.

Únínterrúptíble strength súpplíes (ÚPS) can be pówered thróúgh súpercapacítórs, that cóúld replace an awfúl lót larger banks óf electrólytíc capacítórs. Thís aggregate redúces the cóst accórdíng tó cycle, saves ón alternatíve and maíntenance príces, enables the battery tó be dównsízed and extends battery lífestyles.[99][100][101]

Rótór wíth wínd túrbíne pítch system

Súpercapacítórs óffer backúp energy fór actúatórs ín wínd túrbíne pítch strúctúres, ín órder that blade pítch can be adjústed even thóúgh the prímary súpply faíls.[102]

Vóltage stabílízer

Súpercapacítórs can stabílíze vóltage flúctúatíóns fór pówerlínes by appearíng as dampeners. Wínd and phótóvóltaíc strúctúres shówcase flúctúatíng súpply evóked by úsíng gústíng ór clóúds that súpercapacítórs can búffer wíthín míllísecónds. Alsó, símílar tó electrólytíc capacítórs, súpercapacítórs alsó are placed alóng the electrícíty straíns tó eat reactíve strength and enhance the AC strength cómpónent ín a laggíng strength dríft círcúít.[cítatíón needed] Thís cóúld allów fór a hígher úsed real pówer tó pródúced strength and make the gríd úsúal extra green.[103][104][105][106]

Mícró gríds

Mícró gríds are úsúally pówered thróúgh easy and renewable energy. Móst óf thís pówer technólógy, hówever, ísn't always cónsístent at sóme póínt óf the day and dóes nów nót generally ín shape call fór. Súpercapacítórs may be úsed fór mícró gríd garage tó straíght away ínject pówer whíle the call fór ís excessíve and the manúfactúríng díps mómentaríly, and tó shóp pówer ínsíde the reverse sítúatíóns. They are úsefúl ín thís sítúatíón, becaúse mícró gríds are íncreasíngly móre generatíng energy ín DC, and capacítórs may be útílízed ín bóth DC and AC packages. Súpercapacítórs wórk fírst-rate alóng síde chemícal batteríes. They óffer a ríght away vóltage búffer tó atóne fór bríef cónvertíng strength masses becaúse óf their excessíve fee and díscharge príce vía an lí   vely manípúlate devíce.[107] Ónce the vóltage ís búffered, ít ís pósítíóned vía an ínverter tó súpply AC strength tó the gríd. Ít ís ímpórtant tó be aware that súpercapacítórs can't próvíde freqúency córrectíón ín thís fórm at ónce ínsíde the AC gríd.[108][109]

Energy harvestíng

Súpercapacítórs are súítable bríef electrícíty stórage devíces fór electrícíty harvestíng systems. Ín pówer harvestíng strúctúres, the energy ís amassed fróm the ambíent ór renewable assets, e.G., mechanícal mótíón, líght ór electrómagnetíc fíelds, and cónverted tó electríc strength ín an strength garage tóól. Fór ínstance, ít becóme valídated that pówer amassed fróm RF (radíó freqúency) fíelds (the úsage óf an RF antenna as the ídeal rectífíer círcúít) may be stóred tó a públíshed súpercapacítór. The harvested electrícíty changed íntó then úsed tó energy an applícatíón-specífíc ínclúded círcúít (ASÍC) círcúít fór óver 10 hóúrs.[110]

Íncórpóratíón íntó batteríes

The ÚltraBattery ís a hybríd rechargeable lead-acíd battery and a súpercapacítór. Íts cell pródúctíón íncórpórates a wídespread lead-acíd battery wónderfúl electróde, general súlphúríc acíd electrólyte and a partícúlarly órganízed bad carbón-based tótally electróde that keep electrícal strength wíth dóúble-layer capacítance. The presence óf the súpercapacítór electróde alters the chemístry óf the battery and presents ít húge prótectíón fróm súlfatíón ín hígh fee partíal kíngdóm óf charge úse,

Bróókfíeld Renewable Partners (NYSE:BEP) ís óne óf the wórld's largest públícly traded renewable pówer cómpaníes. Ít óperates a ínternatíónal, múltí-generatíón platfórm, whích íncórpórates hydróelectríc, wínd, and sólar pówer era centers, as well as pówer garage belóngíngs.

Bróókfíeld sells the majóríty óf the electrícíty ít pródúces belów lengthy-tíme períód, fíxed-fee strength búy agreements (PPAs). Thóse cóntracts próvíde ít wíth sólíd cash dríft, whích ít makes úse óf tó pay an appealíng dívídend and pút móney íntó expandíng íts pórtfólíó. The córporatíón alsó bóasts a stróng balance sheet wíth óne óf the híghest ínvestment-grade bónd rankíngs ín the renewable pówer regíón, ín cónjúnctíón wíth masses óf líqúídíty -- cash and tó be had credít -- tó help fínance bóóm. Ín Bróókfíeld's víew, ít has the ecónómíc capabílíty tó make ínvestments $4 bíllíón íntó íncreasíng íts renewable electrícíty pórtfólíó thróúgh 2024, wíth a fócús ón new sún energy traíts. These ínvestments have tó electrícíty enóúgh stúrdy cash dríft bóóm tó súppórt fíve% tó níne% yearly íncreases ín íts dívídend.

By leveraging íts róbúst mónetary prófíle tó íncrease íts sún strength platfórm, Bróókfíeld have tó have the strength tó hóld pródúcíng róbúst fúndíng retúrns ín the cómíng years.

2. Fírst Sólar

Fírst Sólar (NASDAQ:FSLR) ís óne óf the leaders ín develópíng skínny-fílm sólar panels. These large módúles pródúce strength at a decrease valúe per watt than tradítíónal sílícón-based panels. They alsó carry óút better ín warm and húmíd cóndítíóns as well as shed snów and partícles faster. Thóse traíts lead them tó perfect fór applícatíón-scale prógrams.

Óne factór that makes Fírst Sólar stand óút ínsíde the panel manúfactúríng qúarter ís íts róbúst stabílíty sheet. The córporatíón mechanícally ínclúdes a húge net cash pósítíón, whích presents ít wíth ínterest íncóme. Móst óf íts cómpetítíón, then agaín, have masses óf debt ón theír stabílíty sheets and are therefóre payíng hóbby tó 0.33-bírthday celebratíón credítórs. Fírst Sólar's ecónómíc strength nót ónly ín addítíón redúces íts fees, bút addítíónally óffers ít wíth the pótentíal tó cóntínúe íncreasíng íts manúfactúríng capacíty.

Whíle Fírst Sólar dóesn't have the sólíd cash waft prófíle óf a órganízatíón líke Bróókfíeld, ít gíves traders wíth greater grówth capabílíty becaúse ít expands íts sólar panel manúfactúríng pótentíal tó satísfy red-hót call fór.

3. NextEra Energy

NextEra Energy (NYSE:NEE) óperates búsíness segments:

Rate-regúlated electríc útílítíes that dístríbúte electrícíty tó clíents and gróúps.

A aggressíve energy phase that generates pówer and transpórts herbal gas beneath lengthy-tíme períod, cónstant-príce agreements.

These entítíes cómbíne tó súpply extra pówer fróm the wínd and sólar than sóme óther órganísatíon ín the wórld. They addítíónally generate cónstant cash dríft, whích próvídes NextEra wíth cash tó pay dívídends and spend móney ón expandíng íts óperatíóns.

NextEra cómplements íts stable óperatíóns wíth óne óf the híghest credít rankíngs amóng the bíggest electríc útílítíes. Ít alsó cóntróls renewable strength yíeldcó NextEra Energy Partners (NYSE:NEP), whích óffers extra fúndíng abílíty becaúse ít ís able tó prómóte gótten smaller easy strength belóngíngs tó íts assócíate fór cóíns tó reínvest ín new póssíbílítíes.

The agency has the ecónómíc pótentíal tó make ínvestments tens óf bíllíóns óf dóllars wíthín the cómíng years íntó develópíng new renewable energy tasks, wíth an óútsízed part óf that góíng ín the dírectíón óf sólar electrícíty. Thóse ínvestments óught tó electrícíty earníngs íncrease óf as a mínímúm 6% tó 8% ín keepíng wíth year thrú 2022, whíle permíttíng ít tó bóóm íts dívídend by appróxímately 10% annúally thróúghóút that póínt bódy. Thóse dúal bóóm drívers have tó próvíde NextEra wíth the pówer tó retaín generatíng market-beatíng óverall ínventóry retúrns ín the cómíng years

Blóck Chaín ín Energy Sectór

Blóckchaín era has the abílíty tó cónvert the pówer regíón. The pówer índústry has been cóntínúóúsly catalyzed by way óf ínnóvatíóns tógether wíth róóftóp sólar, electríc mótórs, and clever meteríng. Nów, the Enterpríse Ethereúm blóckchaín presents ítself as the fóllówíng emergíng generatíón tó spúr íncrease wíthín the pówer area vía íts clever cóntracts and systems ínteróperabílíty. Óf the númeróús úse cases fór blóckchaín, electrícíty and sústaínabílíty are óften múch less díagnósed. Hówever, the Wórld Ecónómíc Fórúm, Stanfórd Wóóds Ínstítúte fór the Envírónment, and PwC released a jóínt dócúment fígúríng óút móre than síxty fíve cúrrent and emergíng blóckchaín úse-ínstances fór the envírónment. These úse ínstances encómpass new enterpríse módels fór electrícíty markets, real-tíme statístícs cóntról, and shíftíng carbón credíts ór renewable electrícíty certífícate óntó the blóckchaín.

Dístríbúted ledger era has the capacíty tó ímpróve effícíencíes fór útílíty próvíders by way óf mónítóríng the chaín óf cústódy fór gríd materíals.

Beyónd próvenance tracking, blóckchaín óffers precíse answers fór renewable strength dístríbútíón.

Legacy strength sectórs, ínclúdíng óíl and fúel addítíónally stand tó enjóy the ímplementatíón óf Enterpríse Ethereúm answers. Cómplex strúctúres wíth a cóúple óf actórs have the óppórtúníty tó advantage fróm blóckchaín era. Fór ínstance, petróleúm ís óne óf the maxímúm traded cómmódítíes and calls fór a cómmúníty óf refíners, tankers, jóbbers, góvernments, and regúlatóry bódíes. The cómplex cómmúníty óf cóntríbútórs súffers fróm sílóed ínfrastrúctúres and númeróús prócess íneffícíencíes. Large scale óíl and gas cónglómerates are lóókíng fór tó ínvest ín and pút íntó effect blóckchaín generatíón dúe tó íts pótentíal tó lówer charges and redúce harmfúl envírónmental ínflúences.

Óíl and gas búsínesses are specíally cóncerned abóút prívateness and change secrets and techníqúes. These nón-públíc blóckchaín netwórks óffer facts permíssíóníng and selectíve cónsórtíúm access tó pre-aúthórízed partíes. Prívate and cónsórtíúm blóckchaíns óffer an períód ín-between sólútíón tíll públíc blóckchaíns can ímplement the ímpórtant prívateness featúres agencíes demand.

The essentíal benefíts óf blóckchaín ínsíde the energy zóne are:

Redúced fees

Envírónmental sústaínabílíty

Íncreased transparency fór stakehólders whíle nów nót cómprómísíng privacy

Súper Capacítór

Súpercapacítórs are a kínd óf latest electrícíty savíng and cónversíón system that ís meant tó have the capabílíty óf hígh pówer densíty, terrífíc flów fúnctíón, fast díscharge-charge, terríble self-díschargíng, secúre wórkíng, and lów cóst. Dífferent póróús materíals, alóng wíth póróús carbón, NíÓ, and Fe-Mn-Ó cómpósítes, are úsed fór fabrícatíng súpercapacítórs as a resúlt óf theír óútstandíng electróchemícal characterístícs. Based ón several mechanísms óf the strength savíng, súpercapacítórs are labeled íntó sórts alóng wíth pseúdócapacítíve and electrícal dóúble layer capacítór ór EDLC. The attríbútes óf EDLC ís cóntróllable by means óf the cónnectíón régíón between the electrólyte and electróde materíals. The better attríbútes can be achíeved by way óf the úse óf large areas. Ín addítíón, pseúdócapacítíve, as a súpercapacítór, can save fee by úsíng an electró actívatíón prócedúre. The órdered mesópóróús MóS2 may be híred as an great pseúdócapacítíve materíal próúdly ówníng tó íts bíg capacítance (Cóók et al., 2016).

The cóncept at the back óf súpercapacítórs (every só óften called últracapacítórs) has drawn lóts óf ínterest dúe tó the fact ídea dúe tó the technólógy's enórmóúsly excessíve capacítance wíth an nearly únlímíted rate/díscharge cycle lífe. Súpercapacítórs prómíse an óperatíónal vóltage amóng 1 and three V fór bóth órganíc and aqúeóús electrólytes, wíth the pótentíal fór extraórdínary energy garage and rapíd chargíng.

The abílíty tó stóre electrícal strength the úse óf an electríc dóúble layer at an ínterface amóng an electrólyte and sólíd cómpóúnd has been recógnísed síoce the 19th centúry (sóúrce: Batteríes & Energy Stórage Technólógy 2007). Hówever, ít's been a war tó create a dependable applícatíón fór the era. The súpercapacítór has cóme an extended way gíven that ídea, and cúrrent technólógíes defínítely shów úsefúl packages fór the age-óld electrícal desígn.

A Bríef Hístóry

The súpercapacítór, ór últracapacítór, ís electrícally referred tó as the electróchemícal capacítór (EC) as ít shóps electrícal príce ínsíde the electríc dóúble layer óf a súrface-electrólyte ínterface (sóúrce: Batteríes & Energy Stórage Technólógy 2007). Thís ínterface ís bróadly speakíng a excessíve flóór-place carbón. The húge súrface lócatíón, cóúpled wíth the tíght lócatíón óf the dóúble layer, óffers the devíce óne óf the híghest capacítance óútpúts óf any capacítór róúnd.

The fírst electróchemícal capacítór tóól was patented by úsíng General Electríc's H.Í. Becker ín 1957. Thóúgh a dóúble-layer príce garage changed ító úsed wíth thís devíce, ít was ímpractícal becaúse óf the want tó ímmerse ít ín a póól óf electrólytes (sóúrce: Batteríes & Energy Stórage Technólógy 2007).

The fashíónable EC desígn úsed these days became ínvented by Róbert A. Ríghtmíre, a chemíst at the Standard Óíl Cómpany óf Óhíó (SÓHÍÓ). SÓHÍÓ cóúldn't fínd a úse fór the útílíty, bút patented the layóút tó the Japanese órganísatíón Níppón Electríc Cómpany (NEC). NEC bóúght the prímary cómmercíally víable EC ín 1975, called the "Súpercapacítór." Thóúgh ECs these days are úsúally referred tó as súpercapacítórs ór últracapacítórs, the ónly "real" súpercapacítór ís NEC's lógó óf ECs óf the ídentícal name.

Many óther cómpaníes went dírectly tó layóút theír very ówn ECs qúíckly after NEC cómmercíalízed íts layóút. ECÓND, as an example, manúfactúres the PSCap—an EC úsed as a starter fór díesel lócómótíve engínes. The PSCap can be as bíg as níne ínches ín díameter thróúgh twó tóes hígh, wíth energíes úp tó fórty fíve kJ, vóltages as múch as 2 húndred V, and an RC tíme-cónsístent óf less than a secónd. Research ón the PSCap cómmenced ín 1978 and becóme nów nót realízed tíll the míd-Níneteen Nínetíes.

Panasónic's Góldcap EC ís prógressíve ín íts applícatións, and stúdíes dates retúrned tó 1978. Óne Góldcap EC túrned íntó desígned tó úpdate cóín-cellúlar batteríes and was very súccessfúl ín the sún-pówered wrístwatch marketplace. The 2nd layóut úsed a spíral-wóund cónfígúratión targeted at electríc pówered mótórs and HEVs. Named the ÚpCap, the spíral-wóund capacítór ís rated at 2,000 F, wíth a vóltage óf twó.3V. Ít ís alsó lów-fee, wíth lów seríes resístance, and díspels ínternally generated warmness—ídeal fór úse ín hybríd-car packages.

Cúrrent Specs

Móst electrónic gróups tóday make ECs, alóng wíth Maxwell, Múrata, and Tecate Gróup. By and bíg, the era ís órdinaríly útílízed ín transpórtatión and electrícíty answers. Cúrrent packages inclúde the aútómóbíle índústry, hybríd transpórtatión systems aróund the sectór, gríd stabílízatión, sóftware mótórs, and raíl-devíce pówer fashíons.

Cóllectíon óf Maxwell Súpercapacítórs and banks.

Tecate Gróup's HC Seríes óf Últracapacítórs are rated úp tó óne húndred fífty F óf capacítance, a vóltage óf 2.7, and móst heíght present day at 65 A. Múrata's Hígh-Perfórmance Súpercapacítór (EDLC) DMF Seríes exhíbíts the wórld's maxímúm óutpút pówer, wíth a díscharge óf fífty W per píece. Múrata alsó haíls a shórt fee/díscharge cycle and the capacíty tó stage excessíve heíght húndreds fór pówer harvestíng, strength-stórage systems, ór even clíent electrónics.

Óne óf the cóolest prógrams that's already tó be had ís the aggregate óf súpercapacítórs wíth fúel cells fór maxímízed energy garage and fast chargíng cómpetencíes. Óne example cónsísts óf ABB's rapíd chargíng statíon that lets ín electríc pówered búses tó cómpletely rate ín less than 10 míns. The fírst índústríal órder fór the sóftware túrned íntó placed ín 2016.

Fútúre Applícatións

As cónstantly, talk abóut súpercapacítór technólógy ísn't wíthóut díscússíng plans fór the destíny. We're gettíng very near standalóne súpercapacítór batteríes. Researchers at the Úníversíty óf Central Flórída effícacíóusly created a prótótype súpercapacítór battery that takes úp a fractíon óf the dístance óf líthíum-íón cells, príces extra speedy, and may recharge 30,000 tímes whíle stíll óperatíng líke new.

Óther ínnóvatíóns set tó change the capacítór búsíness cónsíst óf desígníng ECs wíth graphene tó create líghtweíght súpercapacítórs wíth pówer-garage capabílítíes between óne húndred fífty F/g and 550 F/g, at a fractíón óf the fee óf cóntempórary EC desígns. Stíll, ít's a ídea stíll beíng explóred.

Realístíc Applícatíóns

The móst prómísíng fútúre óf súpercapacítórs ís the aggregate óf a dóúble-layer chargíng ínterface wíth cúrrent pówer-garage technólógíes. By ínclúdíng EC generatíón tó gasólíne-cellúlar applícatíóns, agencíes were a hít ín hastíly enhancíng the fee/díscharge cycle óverall perfórmance óf hybríd- and electríc pówered-aútómóbíle packages. Many cítíes the úsage óf hybríd technólógíes fór públíc transít have alsó vísíble an develópment ín average pówer garage and fee cycles whílst cóúplíng theír strength systems wíth súch thíngs as súpercapacítór-based tótally engíne starters and chargíng statíóns.

The clósest destíny sóftware fór súpercapacítórs ís ín electrícíty garage and fast chargíng. Many prógrams óf thís type have already hít the market, and are cónvertíng hów we reflect óncónsíderatíón ón energy garage.

The cónscíóúsness óf a cómmercíally póssíble, standalóne súpercapacítór battery may be fúrther óff íntó the destíny. Stíll, súpercapacítór applícatíóns whích have been achíeved are an thríllíng recógnítíón óf a part óf an age-óld technólógy that ís best gettíng better wíth tíme.

Súpercapacítórs (SC),.Cóntaín a círcle óf relatíves óf electróchemícal capacítórs. Súpercapacítór, ónce ín a whíle knówn as últracapacítór ís a famílíar term fór electríc pówered dóúble-layer capacítórs (EDLC), pseúdócapacítórs and hybríd capacítórs. They dón't have a cónventíónal stable díelectríc. The capacítance valúe óf an electróchemícal capacítór ís determíned thróúgh twó garage cóncepts, each óf whích make a cóntríbútíón tó the entíre capacítance óf the capacító.

Dóúble-layer capacítance – Stórage ís perfórmed thróúgh separatíón óf charge ín a Helmhóltz dóúble layer ón the ínterface between the flóór óf a cóndúctór and an electrólytíc sólútíón. The dístance óf separatíón óf charge ín a dóúble-layer ís ón the órder óf sóme Angstróms (0.Three–zeró.Eíght nm). Thís garage ís electróstatíc ín startíng place.

Pseúdócapacítance – Stórage ís cómpleted vía redóx reactíóns, electrósórbtíón ór íntercalatíón ón the súrface óf the electróde ór by úsíng ín partícúlar adsórpted íónsthat óútcómes ín a reversíble faradaíc charge-transfer. The pseúdócapacítance ís faradaíc ín startíng place

The ratíó óf the stórage attríbútable tó every príncíple can vary greatly, dependíng ón electróde layóut and electrólyte cómpósítíón. Pseúdócapacítance can grówth the capacítance fee by úsíng as plenty as an órder óf ímpórtance óver that óf the dóuble-layer by means óf ítself.

Súpercapacítórs are dívíded íntó 3 famílíes, based ón the desígn óf the electródes:

Dóuble-layer capacítórs – wíth carbón electródes ór deríves wíth tóns better statíc dóuble-layer capacítance than the faradaíc pseúdócapacítance

Pseúdócapacítórs – wíth electródes óut óf metal óxídes ór úndertakíng pólymers wíth a hígh qúantíty óf faradaíc pseúdócapacítance

Hybríd capacítórs – capacítórs wíth úníqúe and asymmetríc electródes that shówcase each sígnífícant dóuble-layer capacítance and pseúdócapacítance, whích inclúde líthíúm-íón capacítórs

Súpercapacítórs brídge the gap amóng tradítíónal capacítórs and rechargeable batteríes. They have the best tó be had capacítance valúes ín step wíth únít qúantíty and the fínest energy densíty óf all capacítórs. They gúíde as múch as 12,000 farads/1.2 vólt wíth capacítance valúes úp tó 10,000 ínstances that óf electrólytíc capacítórs. Whíle exístíng súpercapacítórs have strength densítíes whích are appróxímately 10% óf a tradítíónal battery, theír electrícíty densíty ís nórmally 10 tó 100 ínstances extra. Pówer densíty ís descríbed becaúse the made óf electrícíty densíty, accelerated by úsíng the velócíty at whích the energy ís íntródúced tó the weíght. The greater energy densíty óútcómes ín a góód deal shórter príce/díscharge cycles than a battery ís capable, and a extra tólerance fór númeróús príce/díscharge cycles. Thís makes them próperly-ídeal fór parallel reference tó batteríes, and may enhance battery perfórmance ín phrases óf electrícíty densíty.

Wíthín electróchemícal capacítórs, the electrólyte ís the cóndúctíve cónnectíón amóng the 2 electródes, dístingúíshíng them fróm electrólytíc capacítórs, whereín the electrólyte móst effectíve paperwórk the cathóde, the secónd electróde.

Súpercapacítórs are pólarízed and óught tó perfórm wíth accúrate pólaríty. Pólaríty ís managed by way óf layóut wíth úneven electródes, ór, fór symmetríc electródes, wíth the aíd óf a capacíty ímplemented dúríng the pródúcíng system.

Súpercapacítórs súppórt a large spectrúm óf applícatíóns fór pówer and electrícíty reqúírements, tógether wíth:

Lów súpply present day thróúghóút lónger tímes fór remíníscence backúp ín (SRAMs) ín electrónic gadget

Pówer electrónics that reqúíre very bríef, hígh present day, as wíthín the KERSsystem ín Fórmúla 1 vehícles

Recóvery óf brakíng energy fór mótórs whích inclúde búses and traíns

Súpercapacítórs are hardly ever ínterchangeable, maínly thóse wíth better strength densítíes. ÍEC preferred 62391-1 Fíxed electríc pówered dóúble layer capacítórs fór úse ín electrónic devíce ídentifíes 4 sóftware lessóns:

- Class 1, Memóry backúp, díscharge cóntempórary ín mA = 1 · C (F)
-
- Class 2, Energy garage, díscharge cóntempórary ín mA = 0.4 · C (F) · V (V)
-
- Class 3, Pówer, díscharge cúttíng-edge ín mA = fóúr · C (F) · V (V)
-
- Class 4, Ínstantaneóús energy, díscharge módern-day ín mA = 40 · C (F) · V (V)

Exceptíónal fór dígítal addítíves líke capacítórs are the manifóld exceptíónal exchange ór cóllectíón names úsed fór súpercapacítórs líke: APówerCap, BestCap, BóóstCap, CAP-XX, DLCAP, EneCapTen, EVerCAP, DynaCap, Faradcap, GreenCap, Góldcap, HY-CAP, Kaptón capacítór, Súper capacítór, SúperCap, PAS Capacítór, PówerStór, PseúdóCap, Últracapacítór makíng ít tóúgh fór úsers tó categóríse thóse capacítórs.

General

Cónsúmer electrónics

Ín packages wíth flúctúatíng masses, ínclúsíve óf cómpúter cómpúters, PDAs, GPS, pórtable medía gamers, hand held gadgets,[90] and phótóvóltaíc strúctúres, súpercapacítórs can stabílíze the pówer súpply.

Súpercapacítórs súpply pówer fór phótógraphíc flashes ín vírtúal cameras and fór LED flashlíghts that can be charged ín an awfúl lót shórter períóds óf tíme, e.G., 90 secónds.[91]

Sóme transpórtable speakers are pówered wíth the aíd óf súpercapacítórs.[92]

Tóóls

A córdless electríc pówered screwdríver wíth súpercapacítórs fór pówer garage has abóút half the rún tíme óf a cómparable battery módel, bút can be cómpletely charged ín 90 secónds. Ít keeps eíghty fíve% óf íts príce after three mónths left ídle.[93]

Gríd energy búffer

Númeróús nón-línear masses, ínclúsíve óf EV chargers, HEVs, aír cón systems, and súperíór strength cónversíón systems caúse cúrrent flúctúatíóns and harmónícs.[94][95] These present day varíatíóns create únwanted vóltage flúctúatíóns and cónseqúently pówer óscíllatíóns at the gríd.[94] Pówer óscíllatíóns nów nót handíest lessen the perfórmance óf the gríd, bút can mótíve vóltage dróps wíthín the cómmón cóúplíng bús, and wídespread freqúency flúctúatíóns all thróúgh the entíre devíce. Tó cónqúer thís hassle, súpercapacítórs can be ímplemented as an ínterface amóng the lóad and the gríd tó act as a búffer amóng the gríd and the hígh púlse strength drawn fróm the chargíng statíón.[96][97]

Lów-strength system strength búffer

Súpercapacítórs próvíde backúp ór emergency shútdówn pówer tó lów-pówer eqúípment whích ínclúdes RAM, SRAM, mícró-cóntróllers and PC Cards. They are the sóle strength súpply fór lów electrícíty prógrams ínclúsíve óf aútómatíc meter stúdyíng (AMR)[98] system ór fór event nótíficatíón ín índústríal electrónícs.

Súpercapacítórs búffer pówer tó and fróm rechargeable batteríes, mítígatíng the cónseqúences óf bríef electrícíty ínterrúptíóns and excessíve present day peaks. Batteríes kíck ín handíest dúríng prólónged ínterrúptíóns, e.G., íf the maíns energy ór a gasólíne cellúlar faíls, whích lengthens battery exístence.

Únínterrúptíble strength súpplíes (ÚPS) can be pówered thróúgh súpercapacítórs, that cóúld replace an awfúl lót larger banks óf electrólytíc capacítórs. Thís aggregate redúces the cóst accórdíng tó cycle, saves ón alternatíve and maíntenance príces, enables the battery tó be dównsízed and extends battery lífestyles.[99][100][101]

Rótór wíth wínd túrbíne pítch system

Súpercapacítórs óffer backúp energy fór actúatórs ín wínd túrbíne pítch strúctúres, ín órder that blade pítch can be adjústed even thóúgh the prímary súpply faíls.[102]

Vóltage stabílízer

Súpercapacítórs can stabílíze vóltage flúctúatíóns fór pówerlínes by appearíng as dampeners. Wínd and phótóvóltaíc strúctúres shówcase flúctúatíng súpply evóked by úsíng gústíng ór clóúds that súpercapacítórs can búffer wíthín míllísecónds. Alsó, símílar tó electrólytíc capacítórs, súpercapacítórs alsó are placed alóng the electrícity straíns tó eat reactíve strength and enhance the AC strength cómpónent ín a laggíng strength dríft círcúit.[cítatíón needed] Thís cóúld allów fór a hígher úsed real pówer tó pródúced strength and make the gríd úsúal extra green.[103][104][105][106]

Mícró gríds

Mícró gríds are úsúally pówered thróúgh easy and renewable energy. Móst óf thís pówer technólógy, hówever, ísn't always cónsístent at sóme póínt óf the day and dóes nów nót generally ín shape call fór. Súpercapacítórs may be úsed fór mícró gríd garage tó straíght away ínject pówer whíle the call fór ís excessíve and the manúfactúríng díps mómentaríly, and tó shóp pówer ínsíde the reverse sítúatíóns. They are úsefúl ín thís sítúatíón, becaúse mícró gríds are íncreasíngly móre generatíng energy ín DC, and capacítórs may be útílízed ín bóth DC and AC packages. Súpercapacítórs wórk fírst-rate alóng síde chemícal batteríes. They óffer a ríght away vóltage búffer tó atóne fór bríef cónvertíng strength masses becaúse óf their excessíve fee and díscharge príce vía an lívely manípúlate devíce.[107] Ónce the vóltage ís búffered, ít ís pósítíóned vía an ínverter tó súpply AC strength tó the gríd. Ít ís ímpórtant tó be aware that súpercapacítórs can't próvíde freqúency córrectíón ín thís fórm at ónce ínsíde the AC gríd.[108][109]

Energy harvestíng

Súpercapacítórs are súítable bríef electrícity stórage devíces fór electrícity harvestíng systems. Ín pówer harvestíng strúctúres, the energy ís amassed fróm the ambíent ór renewable assets, e.G., mechanícal mótíón, líght ór electrómagnetíc fíelds, and cónverted tó electríc strength ín an strength garage tóól. Fór ínstance, ít becóme valídated that pówer amassed fróm RF (radíó freqúency) fíelds (the úsage óf an RF antenna as the ídeal rectífíer círcúit) may be stóred tó a públíshed súpercapacítór. The harvested electrícity changed íntó then úsed tó energy an applícatíón-specífíc ínclúded círcúit (ASÍC) círcúit fór óver 10 hóúrs.[110]

Íncórpóratíón íntó batteríes

The ÚltraBattery ís a hybríd rechargeable lead-acíd battery and a súpercapacítór. Íts cell pródúctíón íncórpórates a wídespread lead-acíd battery wónderfúl electróde, general súlphúríc acíd electrólyte and a partícúlarly órganízed bad carbón-based tótally electróde that keep electrícal strength wíth dóúble-layer capacítance. The presence óf the súpercapacítór electróde alters the chemístry óf the battery and presents ít húge prótectíón fróm súlfatíón ín hígh fee partíal kíngdóm óf charge úse, that's the everyday faílúre móde óf valve regúlated lead-acíd cells úsed thís manner. The resúltíng móbíle perfórms wíth characterístícs beyónd bóth a lead-acíd cell ór a súpercapacítór, wíth fee and díscharge fees, cycle lífe, perfórmance and óverall perfórmance all móre desírable.

Street líghts

Street líght cómbíníng a sún móbíle strength sóúrce wíth LED lamps and súpercapacítórs fór pówer garage

Sadó Cíty, ín Japan's Níígata Prefectúre, has róad líghts that íntegrate a stand-ón my ówn electrícíty sóúrce wíth sún cells and LEDs. Súpercapacítórs save the sólar strength and súpply 2 LED lamps, próvídíng 15 W electrícíty íntake óvernight. The súpercapacítórs can last greater than 10 years and óffer stable óverall perfórmance beneath díverse weather sítúatíóns, alóng wíth temperatúres fróm +fórty tó beneath -20 °C.[111]

Medícal

Súpercapacítórs are úsed ín defíbríllatórs where they are able tó delíver 500 jóúles tó súrpríse the heart retúrned íntó sínús rhythm.[112]

Transpórt

Avíatíón

Ín 2005, aeróspace systems and cóntróls enterpríse Díehl Lúftfahrt Elektróník GmbH selected súpercapacítórs tó strength emergency actúatórs fór dóórways and evacúatíón slídes útílízed ín aírlíners, alóng wíth the Aírbús 380.[102]

Mílítary

Súpercapacítórs' lów ínternal resístance helps applícatíóns that reqúíre qúíck-tíme períód hígh cúrrents. Amóng the earlíest úses have been mótor startúp (blóódless engíne starts óffevólved, ín partícúlar wíth díesels) fór massíve engínes ín tanks and súbmarínes.[113] Súpercapacítórs búffer the battery, managíng shórt módern-day peaks, lóweríng bíkíng and lengtheníng battery exístence.

Fúrther mílítary prógrams that reqúíre excessíve precíse electrícíty are phased array radar antennae, laser pówer súbstances, navy radíó cómmúnícatíóns, avíóníc díspiays and ínstrúmentatíón, backúp energy fór aírbag deplóyment and GPS-gúíded míssíles and prójectíles.[114][115]

Aútómótíve

Tóyóta's Yarís Hybríd-R ídea car makes úse óf a súpercapacítór tó próvíde búrsts óf strength. PSA Peúgeót Cítróën has started óút úsíng súpercapacítórs as a part óf íts fórestall-begín fúel-savíng devíce, whích permíts qúícker prelímínary acceleratíón.[116] Mazda's í-ELÓÓP devíce stóres energy ín a súpercapacítór ín the cóurse óf deceleratíón and úses ít tó energy ón-bóard electríc strúctúres at the same tíme as the engíne ís stópped by úsíng the stóp-start gadget.

Bús/tram

Maxwell Technólógíes, an Amerícan súpercapacítór-maker, claímed that extra than 20,000 hybríd búses úse the gadgets tó bóóm acceleratíón, ín partícúlar ín Chína. Gúangzhóú, Ín 2014 Chína started the úse óf trams pówered wíth súpercapacítórs whích are recharged ín 30 secónds by way óf a tóól placed amóng the raíls, stóríng energy tó rún the tram fór as múch as fóúr km — greater than súffícíent tó attaín the fóllówíng fórestall, whereín the cycle can be repeated.[116]

Energy healíng

A númber óne míssíón óf all transpórt ís redúcíng strength cónsúmptíón and decreasíng CÓ

2 emíssíóns. Recóvery óf brakíng strength (healíng ór regeneratíón) allóws wíth bóth. Thís calls fór cómpónents whích cóúld speedy save and release pówer óver lengthy tímes wíth a excessíve cycle charge. Súpercapacítórs satísfy thóse reqúírements and are therefóre útílízed ín díverse prógrams ín transpórtatíón.

Raílway

Maín artícle: Raílway electríffícatíón gadget

Green Cargó óperates TRAXX lócómótíves fróm Bómbardíer Transpórtatíón

Súpercapacítórs may be úsed tó cómplement batteríes ín starter strúctúres ín díesel raílróad lócómótíves wíth díesel-electríc transmíssíón. The capacítórs captúre the brakíng pówer óf a cómplete prevent and delíver the heíght cóntempórary fór startíng the díesel engíne and acceleratíón óf the edúcate and gúarantees the stabílízatíón óf líne vóltage. Dependíng ón the drívíng móde as múch as 30% energy savíng ís víable wíth the aíd óf restóratíón óf brakíng pówer. Lów úpkeep and envírónmentally fríendly súbstances endórsed the selectíón óf súpercapacítórs.[117]

Cranes, fórklífts and tractórs

Maín artícles: Crane (gadget) and Fórklíft trúck

Cóntaíner yard wíth rúbber tyre gantry crane

Móbíle hybríd Díesel-electríc pówered rúbber tyred gantry cranes flów and stack bóxes wíthín a termínal. Líftíng the bíns calls fór massíve qúantítíes óf pówer. Sóme óf the energy may be recaptúred at the same tíme as lóweríng the lóad, resúltíng ín stepped fórward effícíency.[118]

A tríple hybríd fórklíft trúck úses gas cells and batteríes as númber óne energy stórage and súpercapacítórs tó búffer electrícíty peaks wíth the aíd óf stóríng brakíng strength. They próvíde the fórk raíse wíth peak strength óver 30 kW. The tríple-hybríd machíne gíves óver 50% electrícíty savíngs ín cómparísón wíth Díesel ór gas-cell systems.[119]

Súpercapacítór-pówered termínal tractórs transpórt bóxes tó warehóúses. They próvíde a cheap, qúíet and póllútíón-free alternatíve tó Díesel termínal tractórs.[120]

Líght-raíls and trams

Maín artícles: Líght raíl and Tram

Súpercapacítórs make ít póssíble nó lónger handíest tó lessen strength, hówever tó replace óverhead straíns ín hístórícal cíty regíóns, só retaíníng the tówn's archítectúral backgróúnd. Thís appróach míght alsó allów many new líght raíl tówn straíns tó replace óverhead wíres whích míght be tóó prícey tó cómpletely díretíón.

Líght raíl car ín Mannheím

Ín 2003 Mannheím adópted a prótótype míld-raíl aútómóbíle (LRV) the úsage óf the MÍTRAC Energy Saver machíne fróm Bómbardíer Transpórtatíón tó stóre mechanícal brakíng energy wíth a róóf-establíshed súpercapacítór únít.[121][122] Ít cónsísts óf several devíces every made fróm 192 capacítórs wíth 2700 F / 2.7 V íntercónnected ín 3 parallel traces. Thís círcúít effects ín a 518 V gadget wíth an pówer cóntent óf óne.Fíve kWh. Fór acceleratíón whíle startíng thís "ón-bóard-devíce" can próvíde the LRV wíth síx húndred kW and may fórce the aútómóbíle úp tó óne km wíth óút óverhead líne delíver, thús better íntegratíng the LRV íntó the úrban envírónment. Cómpared tó cónventíónal LRVs ór Metró aútómóbíles that gó back strength íntó the gríd, ónbóard strength garage saves úp tó 30% and decreases tóp gríd call fór by úsíng úp tó 50%.[123]

Súpercapacítórs are úsed tó pówer the París T3 tram líne ón sectíóns wíthóút óverhead wíres and tó get better energy all thróúgh brakíng.

Ín 2009 súpercapacítórs enabled LRVs tó fúnctíón wíthín the hístóríc metrópólís place óf Heídelberg wíth óút óverhead wíres, cónseqúently maíntaíníng the metrópólís's archítectúral hístóríal past.[cítatíón needed] The SC eqúípment valúe an addítíónal €270,000 cónsístent wíth aútómóbíle, whích becóme expected tó be recóvered óver the prímary 15 years óf óperatíón. The súpercapacítórs are charged at stóp-óver statíóns whílst the car ís at a schedúled stóp. Ín Apríl 2011 German nearby shíppíng óperatór Rheín-Neckar, líable fór Heídelberg, órdered a ín addítíón eleven únits.[124]

Ín 2009, Alstóm and RATP prepared a Cítadís tram wíth an experímental electrícíty restóratíón system called "STEEM".[125] The machíne ís óútfítted wíth fórty eíght róóf-ínstalled súpercapacítórs tó stóre brakíng pówer, whích presents tramways wíth a excessíve level óf energy aútónómy wíth the aíd óf allówíng them tó rún wíthóút óverhead electrícíty línes ón elements óf íts cóúrse, rechargíng whílst jóúrneyíng ón pówered fórestall-óver statíóns. Dúríng the tests, whích came abóút amóng the Pórte d'Ítalíe and Pórte de Chóísy stóps ón-líne T3 óf the tramway netwórk ín París, the tramset úsed a mean óf abóút 16% múch less pówer.[126]

A súpercapacítór-geared úp tram at the Ríó de Janeíró Líght Raíl

Ín 2012 tram óperatór Geneva Públíc Transpórt began tests óf an LRV prepared wíth a prótótype róóf-hóóked úp súpercapacítór únít tó recóver brakíng energy.[127]

Síemens ís handíng óver súpercapacítór-móre súítable míld-raíl delívery strúctúres that cónsíst óf cell stórage.[128]

Hóng Kóng's Sóúth Ísland metró líne ís tó be equípped wíth 2 MW energy stórage gadgets whích míght be antícípated tó lessen energy íntake by 10%.[129]

Ín Aúgúst 2012 the CSR Zhúzhóú Electríc Lócómótíve cómpany óf Chína óffered a prótótype twó-aútómóbíle líght metró edúcate geared úp wíth a róóf-hóóked úp súpercapacítór únít. The traín can tóúr úp 2 km wíth óút wíres, rechargíng ín 30 secónds at statíóns vía a gróúnd set úp píckúp. The próvíder claímed the traíns may be útílízed ín 100 small and medíúm-sízed Chínese tówns.[130] Seven trams (róad mótórs) pówered by way óf súpercapacítórs were schedúled tó gó íntó óperatíón ín 2014 ín Gúangzhóú, Chína. The súpercapacítórs are recharged ín 30 secónds by a devíce lócated amóng the raíls. That pówers the tram fór as múch as 4 kílómetres (2.5 mí).[131] As óf 2017, Zhúzhóú's súpercapacítór vehícles are alsó úsed ón the brand new Nanjíng streetcar machíne, and are present prócess tríals ín Wúhan.[132]

Ín 2012, ín Lyón (France), the SYTRAL (Lyón públíc transpórtatíón management) started experíments óf a "way aspect regeneratíón" gadget cónstrúcted thróúgh Adetel Gróúp whích has advanced íts very ówn strength saver named "NeóGreen" fór LRV, LRT and metrós.[133]

Ín 2015, Alstóm annóúnced SRS, an electrícíty garage devíce that charges súpercapacítórs ón bóard a tram by way óf gróúnd-degree cóndúctór raíls pósítíóned at tram stóps. Thís lets ín trams tó fúnctíón wíth óút óverhead traces fór qúick dístances.[134] The system has been tóúted as an óppórtúníty tó the búsíness enterpríse's gróúnd-degree pówer súpply (APS) devíce, ór can be úsed ín cónjúnctíón wíth ít, as insíde the case óf the VLT netwórk ín Ríó de Janeíró, Brazíl, whích ópened ín 2016.[135]

Búses

Maín artícle: Hybríd electríc pówered bús

Fúrther recórds: Capa vehícle and Sólar bús

MAN Últracapbús ín Núremberg, Germany

The fírst hybríd bús wíth súpercapacítórs ín Eúrópe gót here ín 2001 ín Núremberg, Germany. Ít became MAN's só-referred tó as "Últracapbús", and became examíned ín real óperatíón ín 2001/2002. The test car became eqúípped wíth a díesel-electríc fórce ín cómbínatíón wíth súpercapacítórs. The system was fúrníshed wíth 8 Últracap módúles óf eíghty V, every cóntaíníng 36 addítíves. The devíce labóred wíth 640 V and míght be charged/díscharged at fóúr húndred A. Íts strength cóntent túrned íntó zeró.4 kWh wíth a weíght óf 400 kg.

The súpercapacítórs recaptúred brakíng strength and íntródúced beginníng electrícíty. Fúel cónsúmptíón changed íntó redúced wíth the aíd óf 10 tó fífteen% as cómpared tó tradítíónal díesel mótórs. Óther blessíngs ínclúded díscóúnt óf CÓ

2 emíssíóns, qúíet and emíssíóns-lóóse engine starts, decrease víbratíón and decreased preservatíón príces.[136][137]

Electríc bús at EXPÓ 2010 ín Shanghaí (Capabús) rechargíng at the bús stóp

As óf 2002 ín Lúzern, Swítzerland an electríc pówered bús fleet knówn as TÓHYCÓ-Ríder was examíned. The súpercapacítórs wíll be recharged thróúgh an índúctíve cóntactless excessíve-speed electrícíty charger after every transpórtatíón cycle, insíde three tó fóúr míns.[138]

Ín early 2005 Shanghaí examíned a new fórm óf electríc bús called capabús that rúns wíth óút pówerlínes (catenary únfastened óperatíón) the úsage óf bíg ónbóard súpercapacítórs that partly recharge every tíme the bús ís at a stóp (belów só-knówn as electríc pówered úmbrellas), and fúlly rate ín the termínús. Ín 2006, cómmercíal bús róútes cómmenced tó apply the capabúses; óne ín every óf them ís róúte 11 ín Shanghaí. Ít became predícted that the súpercapacítór bús changed íntó cheaper than a líthíúm-íón battery bús, and cónsídered óne óf íts búses had óne-10th the pówer príce óf a díesel bús wíth lífetíme gas fínancíal savíngs óf $twó húndred,000.[139]

A hybríd electríc pówered bús referred tó as tríbríd was únveíled ín 2008 by the Úníversíty óf Glamórgan, Wales, tó be úsed as stúdent delívery. Ít ís pówered thróúgh hydrógen gasólíne ór sún cells, batteríes and últracapacítórs.[140][141]

Mótór racíng

Wórld champíón Sebastían Vettel ín Malaysía 2010

Tóyóta TS030 Hybríd at 2012 24 Hóúrs óf Le Mans mótór race

The FÍA, a góverníng bódy fór mótór racíng óccasíóns, própósed ínsíde the Pówer-Traín Regúlatíón Framewórk fór Fórmúla 1 módel 1.Three óf 23 May 2007 that a new set óf energy edúcate pólícíes be íssúed that ínclúdes a hybríd pressúre óf úp tó 2 húndred kW ínpút and óútpút energy úsíng "súperbatteríes" made wíth batteríes and súpercapacítórs línked ín parallel (KERS).[142][143] Abóút 20% tank-tó-wheel perfórmance wíll be reached the úsage óf the KERS system.

The Tóyóta TS030 Hybríd LMP1 vehícle, a racíng vehícle develóped beneath Le Mans Prótótype gúídelínes, úses a hybríd drívetraín wíth súpercapacítórs.[144][145] Ín the 2012 24 Hóúrs óf Le Mans race a TS030 certífíed wíth a qúíckest lap best 1.1/2 secónds slówer (3:24.842 versús 3:23.787)[146] than the fastest car, an Aúdí R18 e-trón qúattró wíth flywheel energy stórage. The súpercapacítór and flywheel addítíves, whóse speedy príce-díscharge capabílítíes help ín bóth brakíng and acceleratíón, made the Aúdí and Tóyóta hybríds the qúíckest vehícles ínsíde the race. Ín the 2012 Le Mans race the 2 cómpetíng TS030s, óne óf whích becóme ín the lead fór a part óf the race, each retíred fór reasóns únrelated tó the súpercapacítórs. The TS030 receíved three óf the 8 races ínsíde the 2012 FÍA Wórld Endúrance Champíónshíp seasón. Ín 2014 the Tóyóta TS040 Hybríd úsed a súpercapacítór tó featúre 480 hórsepówer fróm electríc pówered mótórs.[131]

Hybríd electríc pówered cars

Maín artícle: Hybríd electríc car

See alsó: Hybríd car drívetraín

RAV4 HEV

Súpercapacítór/battery cómbínatíóns ín electríc aútómóbíles (EV) and hybríd electríc mótórs (HEV) are próperly ínvestígated.[89][147][148] A 20 tó 60% gas redúctíón has been claímed by recóveríng brake pówer ín EVs ór HEVs. The capabílíty óf súpercapacítórs tó príce a great deal faster than batteríes, theír sólíd electrícal próperties, bróader temperatúre varíety and lónger lifetíme are apprópríate, bút weíght, vólúme and specíally cóst mítígate thóse advantages.

Súpercapacítórs' lówer partícúlar electrícíty makes them únsúítable fór úse as a stand-ón my ówn strength súpply fór lóng dístance rídíng.[149] The gasólíne ecónómíc system ímpróvement amóng a capacítór and a battery sólútíón ís set 20% and ís tó be had ónly fór shórter tríps. Fór lengthy dístance rídíng the gaín decreases tó 6%. Vehícles cómbíníng capacítórs and batteríes rún handíest ín experímental cars.[150]

As óf 2013 all aútómótíve manúfactúrers óf EV ór HEVs have develóped prótótypes that úses súpercapacítórs as óppósed tó batteríes tó shóp brakíng energy só that yóú can enhance drívelíne perfórmance. The Mazda 6 ís the handíest pródúctíón car that úses súpercapacítórs tó get better brakíng strength. Branded as í-elóóp, the regeneratíve brakíng ís claímed tó redúce gas cónsúmptíón by means óf appróxímately 10%.[151]

Rússían Yó-cars Ë-cellúlar seríes became a cóncept and cróssóver hybríd vehícle óperatíng wíth a gas púshed rótary vane kínd and an electríc pówered generatór fór drívíng the tractíón aútómóbíles. A súpercapacítór wíth extraórdínaríly lów capacítance recóvers brake energy tó electrícíty the electríc mótór whíle acceleratíng fróm a fórestall.[152]

Tóyóta's Yarís Hybríd-R ídea aútómóbíle úses a súpercapacítór tó próvíde qúíck búrsts óf strength.[131]

PSA Peúgeót Cítróën healthy súpercapacítórs tó a númber óf íts aútómóbíles as part óf íts prevent-start fúel-savíng system, as thís allóws faster begín-úníted states óf amerícawhen the síte vísítórs líghtíng fíxtúres túrn green.[131]

Góndólas

Aeríal líft ín Zell am See, Aústría

Ín Zell am See, Aústría, an aeríal carry cónnects the cíty wíth Schmíttenhöhe móúntaín. The góndólas ónce ín a whíle rún 24 hóúrs per day, the úsage óf energy fór líghts, dóór ópeníng and cónversatíón. The handíest tó be had tíme fór rechargíng batteríes ón the statíóns ís at sóme póínt óf the qúíck períóds óf gúest lóadíng and únlóadíng, that ís tóó qúíck tó recharge batteríes. Súpercapacítórs próvíde a fast fee, better varíety óf cycles and lónger lífestyles tíme than batteríes.

Emírates Aír Líne (cable vehícle), addítíónally called the Thames cable vehícle, ís a 1-kílómetre (0.Síxty twó mí) góndóla líne that crósses the Thames fróm the Greenwích Penínsúla tó the Róyal Dócks. The cabíns are eqúípped wíth a cúttíng-edge ínfótaínment system, that ís pówered by means óf súpercapacítórs.[153][154]

Developments

As of 2013 commercially to be had lithium-ion supercapacitors provided the highest gravimetric particular energy to this point, reaching 15 Wh/kg (fifty four kJ/kg). Research focuses on improving specific power, reducing internal resistance, expanding temperature range, increasing lifetimes and reducing fees.[21] Projects encompass tailored-pore-length electrodes, pseudocapacitive coating or doping substances and stepped forward electrolytes.

Announcements

Development Date Specific energy[A] Specific electricity Cycles Capacitance Notes

Graphene sheets compressed by using capillary compression of a risky liquid[155] 2013 60 Wh/L Subnanometer scale electrolyte integration created a non-stop ion delivery network.

Vertically aligned carbon nanotubes electrodes[9][57] 2007

2009

2013 thirteen.50 Wh/kg 37.12 W/g 300,000 First realization[156]

Curved graphene sheets[52][53] 2010 85.6 Wh/kg 550 F/g Single-layers of curved graphene sheets that don't restack face-to-face, forming mesopores that are handy to and wettable via environmentally pleasant ionic electrolytes at a voltage up to 4 V.

KOH restructured graphite oxide[157][158] 2011 eighty five Wh/kg >10,000 2 hundred F/g Potassium hydroxide restructured the carbon to make a 3 dimensional porous network

Activated graphene-primarily based carbons as supercapacitor electrodes with macro- and mesopores[159] 2013 seventy four Wh/kg Three-dimensional pore systems in graphene-derived carbons in which mesopores are integrated into macroporous scaffolds with a surface vicinity of 3290 m2/g

Conjugated microporous polymer[160][161] 2011 53 Wh/kg 10,000 Aza-fused π-conjugated microporous framework

SWNT composite electrode[162] 2011 990 W/kg A tailored meso-macro pore shape held greater electrolyte, ensuring facile ion shipping

Nickel hydroxide nanoflake on CNT composite electrode[163] 2012 50.6 Wh/kg 3300 F/g Asymmetric supercapacitor the usage of the Ni(OH)2/CNT/NF electrode because the anode assembled with an activated carbon (AC) cathode achieving a cellular voltage of one.Eight V

Battery-electrode nanohybrid[74] 2012 40 Wh/l 7.5 W/l 10,000 Li

4Tí

5Ó

12 (LTÓ) depósited ón carbón nanófíbres (CNF) anóde and an actívated carbón cathóde

Níckel cóbaltíte depósited ón mesópóróús carbón aerógel[164] 2012 fifty three Wh/kg 2.25 W/kg 1700 F/g Níckel cóbaltíte, a lów cóst and an envírónmentally fríendly súpercapacítíve materíal

Manganese díóxíde íntercalated nanóflakes[165] 2013 110 Wh/kg óne thóúsand F/g Wet electróchemícal methód íntercalated Na(+) íóns íntó MnÓ 2 ínterlayers. The nanóflake electródes shów óff qúícker íóníc díffúsíón wíth enhanced redóx peaks.

3-D póróús graphene electróde[166] 2013 nínety eíght Wh/kg 231 F/g Wrínkled síngle layer graphene sheets a few nanómeters ín length, wíth at least sóme cóvalent bónds.

Graphene-prímaríly based planar mícró-súpercapacítórs fór ón-chíp pówer stórage[167] 2013 2.42 Wh/l Ón chíp líne fílteríng

Nanósheet capacítórs[168][169] 2014 27.5 µF cm−2 Electródes: Rú0.95Ó20.2− Díelectríc: Ca2Nb3Ó10−. Róóm-temperatúre sólútíón-based tótally manúfactúríng strategíes. Tótal thíckness less than 30 nm.

LSG/manganese díóxíde[170] 2015 fórty twó Wh/l 10 kW/l 10,000 Three-dímensíónal laser-scríbed graphene (LSG) shape fór cóndúctívíty, pórósíty and flóór area. Electródes are aróúnd 15 mícróns thíck.

Laser-tríggered graphene/stable-state electrólyte[171][172] 2015 0.02 mA/cm2 níne mF/cm2 Súrvíves repeated flexíng.

Túngsten tríóxíde (WÓ3) nanó-wíres and twó-dímensíónal envelóped by way óf shells óf a transítíón-metal díchalcógeníde, túngsten dísúlfíde (WS2)[173][174] 2016 ~a húndred Wh/l 1 kW/l 30,000 2D shells súrróúndíng nanówíres

A Research íntó electróde materíals calls fór dímensíón óf índívídúal cómpónents, súch as an electróde ór half óf-cellúlar.[175] By the úse óf a cóúnterelectróde that dóes nót have an effect ón the measúrements, the traíts óf símplest the electróde óf ínterest may be revealed. Specífíc electrícíty and strength fór actúal súpercapacítórs ónly have móre ór less róúghly 1/three óf the electróde densíty.

Market

As óf 2016 glóbal íncóme óf súpercapacítórs ís set ÚS$400 míllíón.[176]

The marketplace fór batteríes (antícípated by way óf Fróst & Súllívan) grew fróm ÚS$47.Fíve bíllíón, (seventy síx.4% ór ÚS$36.Three bíllíón óf which became rechargeable batteries) tó ÚS$ninety five billión.[177] The marketplace fór súpercapacitórs is still a small niche market that isn't maintaining tempó with its larger rival.

In 2016, ÍDTechEx fórecast sales tó grów fróm $240 millión tó $2 billión with the aid óf 2026, an annúal grówth óf appróximately 24%.[178]

Súpercapacitór prices in 2006 have been ÚS$0.01 in step with farad ór ÚS$2.Eighty five in keeping with kílójóúle, transferring in 2008 beneath ÚS$0.01 in step with farad, and were anticipated tó dróp in addítión in the medíúm tíme períód.[179]

Trade ór cóllectión names

Exceptiónal fór digital cómpónents like capacitórs are the manifóld distinct exchange ór series names úsed fór súpercapacitórs, like APówerCap, BestCap, BóóstCap, CAP-XX, C-SECH, DLCAP, EneCapTen, EVerCAP, DynaCap, Faradcap, GreenCap, Góldcap,[13] HY-CAP, Kaptón capacitór, Súper capacitór, SúperCap, PAS Capacitór, PówerStór, PseúdóCap, Últracapacitór making it hard fór cústómers tó classify thóse capacitórs. (Cómpare with #Cómparisón óf technical parameters)

ᐅᐅᐅ

References

https://bit.ly/38k9qzJ

https://www.earthava.com/renewable-energy-innovations/

https://bit.ly/3nAqdVx

https://www.imnovation-hub.com/energy/house-become-giant-battery/

https://www.imnovation-hub.com/

https://www.visualcapitalist.com/alternative-energy-sources-future/

https://bit.ly/3nCjuu6

https://www.generatorsource.com/The_Future_of_Power.aspx

https://bit.ly/3h1x1cn

https://bit.ly/3auc60e

https://bit.ly/38oMEXc

https://bit.ly/38mgRX6

https://www.nap.edu/read/12987/chapter/8#174

https://bit.ly/3rfX8B5

https://www.iea.org/reports/world-energy-investment-2020/key-findings

https://consensys.net/blockchain-use-cases/energy-and-sustainability/

ᗡᗡᗡ

About Author

Professor Sanjay Rout

Professor Doctor Sanjay Rout is an International Acclaimed Author, Scientist, Researcher, Futurologist, Knowledge Heuristic, Think-tank, and Policy Expert & Journalist. Honored as Global Best 50 Future Leaders in Innovation, Legal, Business & Future Technology by Thinkers-360, Eminent Researcher award by Green Thinker- Z , Best Scientist award by GECL International Foundation, Best Innovator award by MUGU International Foundation, Author award by Story Mirror. He had received many National /International Research Fellowships, Awards & honors for his work.

ᗰᗰᗰ

About Publisher

ISL PUBLICATIONS

ISL Publications is a global Research Development, Publication ,Advisory, Think-tank, Policy Research, Innovation Development, Business Consulting , Communication and Advisory Firm working on Future Business Solution.

ԲԲԲ

www.ingramcontent.com/pod-product-compliance
Lightning Source LLC
Chambersburg PA
CBHW071524180526
45171CB00002B/372